Digital Drive, E-Governance and Internet Services in India

Quality Dimensions

By

Sri Ram Khanna

•

Paramjeet Singh

•

Consumer Voice Research Team
(**Arvinder Kaur, Disha Negi, Malvika Garg** and **Yukti Arora**)

in association with

New Century Publications
New Delhi, India

NEW CENTURY PUBLICATIONS
4800/24, Bharat Ram Road,
Ansari Road, Daryaganj,
New Delhi - 110 002 (India)

Tel.: 011-2324 7798, 4358 7398, 4101 7798
E-mail: indiatax@vsnl.com • info@newcenturypublications.com
www.newcenturypublications.com

Editorial office:
LG – 7, Aakarshan Bhawan,
4754-57/23, Ansari Road, Daryaganj,
New Delhi – 110 002

Tel.: 011-4356 0919

First Published: **2018**

ISBN: **978-81-7708-465-8**

Published by New Century Publications and printed at Milan Enterprises, New Delhi.

Designs: Patch Creative Unit, New Delhi.

PRINTED IN INDIA

About the Book

Digitization has partially or fully replaced many tasks that were previously done by human labourers. At the same time, computers have made some workers much more productive. Another consequence of digitization is that it has drastically reduced the costs of communication among people across different organizations and locations.

'Digital India (DI) Vision 2019' is an initiative of the Government of India to integrate Government Ministries and Departments with the people of India. It aims at ensuring the availability of government services to citizens electronically by reducing paperwork. The initiative also includes plans to connect rural areas with high-speed internet networks. DI has three core components. These include creation of digital infrastructure, delivering services digitally and digital literacy.

Similarly, National e-Governance Plan (NeGP) of the Government of India aims to make all Government services accessible to the common man in his locality, through common service delivery outlets and ensure efficiency, transparency and reliability of such services at affordable costs to realize the basic needs of the common man.

Digital drive and e-governance are both based on internet services. Hence, the quality of internet services is of crucial importance.

The present work incorporates a study—in the form of a comprehensive survey involving 52,000 respondents—to assess the quality of internet services (QoIS) in India. The survey undertaken by VOICE and completed in October 2017 focused on various dimensions of internet services, particularly the QoIS and its impact on users. Based on the findings of the survey, VOICE made a number of recommendations to Telecom Regulatory Authority of India (TRAI) and the Government of India. The work provides deep insights into the structure and regulation of internet services in India and hence is an invaluable source of information for various stakeholders in the field.

About the Authors

Professor Sri Ram Khanna, formerly Dean of the Faculty and Head of the Department of Commerce, Delhi School of Economics (DSE), University of Delhi holds Masters Degree in Commerce, Degree in Law and Doctorate in International Business. Specializing in marketing and international business, he has published 7 books and over 100 research papers in academic journals of repute. He was a Visiting Professor at Texas A&M University, USA (1996). He has also served on the Board of Consumers International, London (2000-2007) and the Life Insurance Corporation of India (2001-2003). He is also the recipient of an award from Indian Merchants Chamber.

Paramjeet Singh holds a Masters Degree in Economics and is a professional researcher and consultant and heads MARCOMRES Solutions Pvt. Ltd., a consultancy firm. He is also Director (Research) and a board member of Consumer Voice, New Delhi.

Arvinder Kaur is currently working as a research associate with Consumer Voice, New Delhi. She holds Ph.D. degree in Economics from Guru Nanak Dev University, Amritsar.

Disha Negi is a post-graduate in commerce from Delhi School of Economics (DSE), University of Delhi. She is presently working as a senior research fellow with Consumer Voice, New Delhi.

Malvika Garg is a post-graduate in international business from Delhi School of Economics (DSE), University of Delhi. Currently, she is working as a senior research fellow with Consumer Voice, New Delhi.

Yukti Arora is a post-graduate in commerce from Delhi School of Economics (DSE), University of Delhi. She was associated with this project as a senior research fellow.

Contents

About the Book ... *v*
About the Authors ... *vi*
Preface ... *xi-xiv*
Abbreviations/Acronyms ... *xv-xvii*

1. Digital Drive in India **1-12**
1.1 Digitization and the Digital Economy
1.2 Digitalization of Banks
1.3 Aadhaar-enabled Payment System (AEPS)
1.4 Digitization of Stock Exchanges
1.5 National Electronic Funds Transfer (NEFT)
1.6 Digital Wallet
1.7 Unified Payment Interface (UPI) of India
1.8 RuPay
1.9 Digital India (DI): Vision 2019

2. E-governance Initiatives **13-23**
2.1 Governance Defined
2.2 What is E-governance?
2.3 National e-Governance plan (NeGP)
2.4 E-governance and Internet Infrastructure
2.5 Unique Identification (UID) System
2.6 Electronic Technology in Tax Administration
2.7 E-commerce
2.8 Electronic Services for Rural Development

3. Importance, Uses and Applications of Internet **24-34**
3.1 Growth of Internet: India and the World
3.2 Motivation and Barriers to Internet Usage
3.3 Impacts of Internet on Day-to-day Life
3.4 Internet and Economy
3.5 Leisure and Recreation

4. Internet and Human Capital Formation 35-48
4.1 Internet for Educational and Skills Development
4.2 Internet for Health and Medical Care
4.3 Influence of Internet on Habits
4.4 Internet for Research and Academics

5. Internet and Social Capital Formation 49-62
5.1 Social Interactions
5.2 Internet and Teenagers
5.3 Search for Partner
5.4 Political Participation
5.5 Internet for People with Special Needs

6. Impact of Internet on Industry, Business and Employment 63-77
6.1 Industry and Services
6.2 Consumer Centricity of Business
6.3 Marketing and Advertising
6.4 Online Shopping
6.5 Travelling Displacement
6.6 Employment and Earnings

7. Umbrella Issues in Internet 78-86
7.1 Impact on Everyday Life
7.2 Gender Inequality
7.3 Privacy
7.4 Internet and Law
7.5 Impact of Mobile Internet

8. VOICE Survey of Quality of Internet Services (QoIS) in India 87-97
8.1 Four Phases of the Growth of Internet in India
8.2 Why is Quality of Internet Service (QoIS) Important?
8.3 Objectives of the Study

8.4 Scope of the Study
8.5 Situational Analysis
8.6 Sampling and Data Collection

9. Time and Money Spent on Main Mode of Internet **98-108**
9.1 Time Spent on Internet: Wireless Service
9.2 Money Spent on Internet: Wireless Service
9.3 Time Spent on Internet: Wireline Service
9.4 Money Spent on Internet: Wireline Service
9.5. Relation between Time and Money

10. Quality of Internet Services **109-135**
10.1 Quality of Service (QoS) Defined
10.2 Quality of Internet Services (QoIS)
10.3 Regulation of QoIS
10.4 Survey on QoIS
10.5 Parameters of Quality of Service
10.6 Overall Quality of Service
10.7 Analytical Comparisons

11. Speed of Internet Services **136-150**
11.1 Consumer Perception about Speed
11.2 QoS Standards for Wireless Data Services
11.3 Promised Speed and Actual Speed
11.4 Speed Performance and Network Strength
11.5 Data Analysis
11.6 Findings and Conclusion
11.7 Recommendations

12. Consumer Problems, Complaints and Redressal Mechanism **151-160**
12.1 Summary of Problems Experienced with Internet Service
12.2 Frequently Occurring Problems (FoPs)
12.3 Conclusion

13. Consumer Detriment: Time Loss and Money Loss 161-165

13.1 Consumer Detriment: Telecom Circle-wise
13.2 Consumer Detriment: Age and Gender-wise
13.3 Consumer Detriment: Education and Profession-wise
13.4 Consumer Detriment: SEC-wise
13.5 Conclusion

14. Recommendations Based on the Findings of VOICE Survey 166-169

14.1 Resume Assessment of QoS through Third Party Assessment
14.2 Compensate for Loss of Time and Money
14.3 Simpler and Easier Consumer Complaint System
14.4 Promote Internet Applications among Rural and Lower Income Consumers
14.5 E-governance
14.6 Mandatory Minimum Download Speed
14.7 Robust and User-friendly TRAI MySpeed app
14.8 Information about Signal Loss

15. Challenges, Risks and Cyber Security 170-186

15.1 Challenges of Expanding Internet Base
15.2 Risks Posed by Internet
15.3 Vulnerability of Cyber Space
15.4 National Cyber Security Policy (NCSP), 2013

References/Bibliography 187-198

Index 199-206

Preface

VOICE is a not-for-profit body dedicated to consumer education and empowerment. It was set up by students and teachers on the Delhi School of Economics campus in 1983 and has since continued its work to protect Indian consumers. It is a member of Consumers International. Most of the studies it takes up are either for guidance of consumers or for policy advocacy. The present study was conceived to focus on the impact of poor quality of internet services on usage of internet for various purposes. It was designed to bring forth all the evidence in regard to the importance of quality of internet service (QoIS) so that the Telecom Regulatory Authority of India (TRAI) and the Government of India could get a comprehensive appreciation of all factors that are concerned with quality of internet service.

There are various services for the consumers which are delivered through internet and consumers are accessing/using government and other services through the internet. The study assesses the impact of quality of internet services delivered to consumers and also lists the shortcomings/impediments which require attention for further enhancing this quality.

The present work is organized into 15 chapters.

Chapter 1 is introductory in nature, highlighting the importance of digital technology in the modern world. It records various initiatives taken in recent years to move India towards a digital economy. It also explains the Digital India (DI): Vision 2019 of the Government of India to integrate Government Ministries/Departments with the people of India.

Chapter 2 explains and examines the role of electronic technology in the delivery of government services and dissemination of information to the citizens. E-governance is an evolutionary phenomenon in India, percolating to almost all segments of the Indian economy. Various initiatives for e-governance are set forth in this chapter.

Chapter 3 spells out the importance of internet in the modern world. More importantly, it records the uses and applications of internet in various walks of life.

Chapter 4 examines the impact of internet on human capital. It covers such issues that shape human lives and enhance the quality thereof such as education and health. Internet also has a profound impact on the formation of habits.

Chapter 5 discusses the impact of internet on social capital, covering such areas as social utilities provided by internet, political participation, and finding a partner. It also describes the online methods of socializing being adopted by teenagers and contribution of internet in helping people with special needs.

Chapter 6 throws light on the role of regulatory bodies, internet service providers and internet infrastructure in creating and maintaining the supply of internet services for industry, business, employment and earnings, and standards of professionalism.

Chapter 7 is devoted to the umbrella issues in internet. Umbrella issues encompass all the areas that have an impact on all aspects of behaviour, as for example, gender inequality and privacy issues.

Chapter 8 contains the broad outline of the survey conducted by VOICE pertaining to the quality of internet service (QoIS) in India. It discusses the objectives, scope and methodology adopted for the survey.

Chapter 9 deals with the *average time spent daily on internet*, and *average expenditure incurred on internet services per month* by the respondents on the basis of mode of internet use (wireless or wireline). The relationship between these two variables has also been examined. It was found that the respondents in rural areas and lower income classes spend relatively more time on an average on their main mode of internet use as compared to others.

Chapter 10 covers the findings of the survey pertaining to quality of internet service in India. The total number of respondents covered during the fieldwork was 52,000, out of which 35,503 (68.3 percent) were those who reported wireless as their main mode of internet use and 16,497 (31.7 percent) reported wireline as their main mode of internet use. The contents of this chapter form the main basis of recommendations made to the Government of India and the Telecom Regulatory Authority of India (TRAI).

Chapter 11 pertains to supplementary study conducted along with the main study of QoIS, i.e. the audit of speed. It gives a brief picture of the actual *internet speeds* of wireless mode available across the country. The chapter concludes on the note that 3G and 4G speeds are defined very loosely, leaving enough room for the service providers to make false claims and misleading advertisements that prompts consumer to make wrong choices.

Chapter 12 sets forth the problems and complaints of the consumers regarding the internet services.

Chapter 13 throws light on the problems and detriment faced by the consumers. Problems encapsulate various issues faced by customers relating to internet services and their providers.

Chapter 14 contains the recommendations made to Government of India and TRAI to improve the quality of internet services.

Chapter 15 is the concluding chapter which discusses the challenges posed by the extending base of internet, risks involved in the use of internet and the cyber security policy of the Government of India.

It has been established that internet has become all-pervasive and has hardly left any segment of the economy untouched. However, the concept of QoIS still remains foreign for the Government, TRAI and service providers. Most consumers are also not able to distinguish between poor, average and good quality of service. A lot has been left to be desired when it comes to getting promised quality of service.

This book is a précis of a larger study "Assessment of the Impact of Quality of Internet Service on Internet Users" which subsumes three volumes. Volume 1 gives introduction to the study, discusses objectives, scope and methodology in detail, besides expounding review of literature and the framework developed through it. It also presents highlights of the study.

Volume 2 is based on the perceptions of internet users on five quality parameters—availability, speed, reliability, billing and charges and help (customer care) services. It also presents the result of 'speed test audit' that was conducted across eight states in all the four telecom circles. Volume 3 assesses the impact of QoIS on different aspects of life of the people who use internet services. To read full report, please visit our blog- https://needqois.wordpress.com.

Acknowledgments

This work was successfully completed with the co-operation, suggestions and hard work on the part of a host of experts and members of the VOICE research team. I put on record my deep appreciation and gratefulness to all of them. In particular, I thank the following scholars and researchers for their invaluable contributions at various stages of the completion of this study: Professor Gurjeet Kaur, Jammu University; Professor H.K. Dangi, Department of Commerce, Delhi School of Economics, Delhi University; Dr. Priyanka Joshi, Narsee Monjee School of Business, Mumbai and Mr. S.C. Saxena, formerly of TRAI. I am also indebted to my co-authors Shri Paramjeet Singh; Dr. Arvinder Kaur; Ms. Malvika Garg; Ms. Yukti Arora; and Ms. Disha Negi.

I gratefully acknowledge the generous grant by the Ford Foundation to VOICE which enabled this study to be undertaken on a national scale. It would not have been possible to undertake this study without this grant.

December 19, 2017

Professor Sri Ram Khanna
Consumer VOICE,
E - 34, East of Kailash,
New Delhi.

Abbreviations/Acronyms

Ads	Advertisements
ANOVA	Analysis of Variance
App	Applications
ASSOCHAM	Associated Chambers of Commerce of India
BHQ	Block Headquarter
B-O-P	Balance of Payments
BTS	Base Transceiver Station
BSNL	Bharat Sanchar Nigam Limited
CATI	Computer Assisted Telephone Interviewing
CCE	Customer Care Executive
CCG	Centre for Communication Governance
CCS	Customer Care Services
CD	Consumer Detriment
CEO	Chief Executive Officer
CMTS	Cellular Mobile Telephone Services
CSCs	Common Service Centres
CSS	Customer Satisfaction Surveys
CUTS	Consumer Unity & Trust Society
DHQ	District Headquarter
DeITY	Department of Electronics and Information Technology
DISNIC	District Information System of the National Informatics Centre
DoT	Department of Telecommunications
E-KYC	E-Know Your Customer
E-WoM	E-Word of Mouth
FCC	Federal Communications Commission
FDI	Foreign Direct Investment
FICCI	Federation of Indian Chambers of Commerce & Industry
F2F	Face to Face
FTS	Fixed Telephone Services
FUP	Fair Usage Policy

GDP	Gross Domestic Product
Govt.	Government
GTPL	Gujarat Telelink Private Limited
HD	High Definition
HSC	Higher Secondary Certificate
ICRIER	Indian Council for Research on International Economic Relations
ICT	Information and Communications Technology
IDI	ICT Development Index
IIM	Importance Influence Matrix
IoT	Internet of Things
IRCTC	Indian Railway Catering and Tourism Corporation
ISP	Internet Service Provider
IT	Information Technology
ITU	International Telecommunications Union
MGNREGA	Mahatma Gandhi National Rural Employment Guarantee Act
Mins.	Minutes
MMP	Mission Mode Projects
MNO	Mobile Network Operator
MOOCs	Massive Open Online Courses
MTNL	Mahanagar Telephone Nigam Limited
NCR	National Capital Region
NeGP	National E-Governance Plan
NGO	Non-governmental Organization
NIC	National Informatics Centre
OHD	Open House Discussion
OOH	Out of Home
PCI	Per Capita Income
QoIS	Quality of Internet Service
QoS	Quality of Service
Rel Comm	Reliance Communications
RTO	Regional Transport Authority
SEC	Socio-economic Classification

SEM	Structural Equation Modelling
SHQ	State Headquarter
SIM	Subscriber Identity Module
SMEs	Small and Medium Enterprises
SPs	Service Providers
SRW	Standard Regression Weights
SSC	Senior Secondary Certificate
TRAI	Telecom Regulatory Authority of India
TSPs	Telecom Service Providers
UIDAI	Unique Identification Authority of India
UK	United Kingdom
US	United States of America
UTs	Union Territories
VOICE	Voluntary Organization in Interest of Consumer Education
VoIP	Voice over Internet Protocol
Y-o-Y	Year-on-Year

1

Digital Drive in India

Around the world, digital technology is seen instrumental in economic development. According to a 2012 World Bank report "a ten percentage point increase in fixed broadband generates a 1.35 percent increase in per capita GDP for developing countries and a 1.19 percent increase for developed countries". This indicates that broadband is much more significant for economic growth in developing countries as compared to developed countries.

The Digital Library of India is a portal with over one million books available free-to-read to anyone with internet connection and a means to display the electronic text. This unlocked access unfolds how internet comes handy in benefiting emerging economies by making large volume of researches available to academic, professionals, public servants and the general population.

Internet as a technology has the potential to narrow down the gap between developed and emerging economies by enabling access, sharing of useful information and by connecting virtual communities of interest. Internet is not passive; rather it drives activity and cannot be ignored for its fast and low-cost alternatives to traditional models of research, business, consumption and communication.

1.1 Digitization and the Digital Economy

Digitization is the process by which technology lowers the costs of storing, sharing, and analyzing data. This process has changed how consumers behave, how industrial activity is organized, and how governments operate. Digitization has coincided with the increased prominence of platforms and marketplaces that connect diverse agents in social and economic activity. Platforms are most readily identified with their technical standards, i.e. engineering specifications for hardware and standards for software. The pricing and product strategies that platforms use differ from those of traditional firms because of the presence of network effects. Network effects are within platforms because participation by one group affects the utility of another group. Furthermore, network effects make the analysis of competition between platforms more complex than the analysis of competition between traditional firms.

Digitization has partially or fully replaced many tasks that were previously done by human labourers. At the same time, computers have made some workers much more productive. Another consequence of digitization is that it has drastically reduced the costs of communication between workers across different organizations and locations. This has led to a change in the geographic and contractual organization of production.

Privacy and data security is an area where digitization has substantially changed the costs and benefits for various economic actors. Traditional policies regarding privacy circumscribed the ability of government agencies to access individual data. However, the large-scale ability of firms to collect, parse, and analyze detailed micro level data about consumers has shifted the policy focus. Now, the concern is whether access to consumer data of firms should be regulated and restricted or not.

There are many other policies related to digitization that are of interest to a cross-section of stakeholders. For example, digitization may affect government effectiveness and accountability. Digitization also makes it easier for firms in one jurisdiction to supply consumers in another. This creates challenges for tax enforcement. Many safety and quality enforcement regulations may no longer be necessary with the advent of online reputation systems. Lastly, digitization is of great importance to health care policy. For example, electronic medical records have the potential to make healthcare more effective but pose challenges to privacy policy.

1.1.1 Digital Economy: It refers to an economy that is based on digital technologies. In this new economy, digital networking and communication infrastructures provide a global platform over which people and organizations devise strategies, interact, communicate, collaborate and search for information. It is widely accepted that the growth of the digital economy has widespread impact on the whole economy. Various attempts at categorising the size of the impact on traditional sectors have been made.

Given its expected broad impact, traditional firms are actively assessing how to respond to the changes brought about by the digital economy. For corporations, timing of their response is of essence. Banks are trying to innovate and use digital tools to improve their traditional business.

The digital drive in India has percolated to almost all the sectors of the economy.

1.2 Digitalization of Banks

Digital transformation of banking services has brought greater transparency in financial services. Digitalization of banks requires various steps such as retail credit processes, digital approaches like mobile application and automation of credit decision. Customers are benefited from higher speed, increase in efficiency and simplicity in banking system. Risk modelling has become possible with digitalization because increased use of digit banking services by customers generate more data. New competitors can operate at much lower costs with improvements in the traditional banking operations, branch network and legacy IT system.

Machine learning techniques in banks have improved the predictability of credit early-warning system. Digitalization of banking operations has reduced paperwork and increased client interaction. Automated monitoring of customers and optimized restructuring solutions have led to minimization of losses. Consumers benefit from faster approvals of credit decisions by banks. Application of advanced-analytics techniques has led to reduction of credit risk costs.

1.2.1 Use of Credit Cards: Digitalization has played an important part in changing consumer behaviour towards credit card debt. Digitalization has increased the usage of credit card for online shopping. Online shopping has become easier and faster because these cards are easy to carry and readily substitute the traditional payment system such as cheques and cash. Consumer attitudes towards credit card are influenced by many economic and demographic factors such as their income, race, age, education and gender. The introduction of credit cards has changed the traditional pattern of consumption smoothing and debt repayment as spending through credit card is not as painful as spending cash.

1.3 Aadhaar-enabled Payment System (AEPS)

Government of India is preparing a platform for realizing a cashless society by developing a common mobile phone app that can be used by shopkeepers and merchants for receiving Aadhaar-enabled payments, bypassing credit and debit cards, PIN and password. The mobile application is used for authenticating biometrics of customers making Aadhaar-enabled payment. The programme will help the Government undertake situations like the recent demonetization and control black money threats while bringing in more financial transaction transparency. People can link their Aadhaar (12-digit number) with their bank accounts and use AEPS for money transfer, balance enquiry, cash deposits or withdrawals and inter-bank transactions. The funds can be directly transferred from the customer's bank accounts to the merchant's. There is a need to create awareness among people about this mode of transaction.

1.4 Digitization of Stock Exchanges

There are 21 recognised stock exchanges in India, including the Over the Counter Exchange of India (OTCEI) for small and new companies, the old established Bombay Stock Exchange (BSE) and the newer National Stock Exchange (NSE) that was set up as a model exchange to provide nationwide services to investors.

The Indian capital market is significant in terms of its degree of development, its volume of trading and of the quality of its automated trading and settlement. The capital markets—though they have suffered from periodic scandals and problems in the past—look much closer in quality to those of developed markets than those in the majority of emerging markets. The range and quality of available securities is evidenced by the substantial flows of foreign institutional investors.

The setting up of the NSE as an electronic trading platform set a benchmark of operating efficiency for other stock exchanges in the country. The establishment of National Securities Depositary Ltd. (NSDL) in 1996 and Central Depository Services (India) Ltd. (CSDL) in 1999 has enabled paperless trading in the exchanges. This has also facilitated instantaneous electronic transfer of securities and eliminated the risks to the investors arising from bad deliveries in the market, delays in share transfer, fake and forged shares and loss of scrips. The electronic funds transfer (EFT) facility combined with dematerialisation of shares has created a conducive environment for reducing the settlement cycle in stock markets.

1.4.1 Dematerialisation (Demat): The concept of demat was introduced in Indian capital market in 1996 with the setting up of NSDL. Demat is the process by which the physical certificates of an investor are converted to an equivalent number of securities in electronic form and credited in the investor's account with his depository participant (DP).

When an investor decides to have his shares in electronic form, he should approach a depository participant (DP)—who is an agent of the depository—and open an account. He should then request for the dematerialisation of certificates by filling up a dematerialisation request form (DRF), which is available with the DP and submitting the same along with the physical certificates.

1.4.2 Rematerialisation: If the investor wishes to get back his securities in physical form, all he has to do is to request his DP for rematerialisation of the same. Rematerialisation is the term used for converting electronic holdings back into certificates. An investor's DP will forward his request to NSDL, after verifying that he has the necessary balance. NSDL in turn will intimate the registrar who will print the certificates and dispatch the same to the investor.

1.5 National Electronic Funds Transfer (NEFT)

Electronic Funds Transfer (EFT) system was introduced in the mid-1990s. EFT facilitates transfer of funds from one bank account to another. The EFT system is currently only permitted to be used for government transactions and RBI-initiated payments. This system is now progressively being replaced with the National Electronic Funds Transfer (NEFT). NEFT is an electronic message-based payment system, and was introduced by the Reserve Bank of India in November 2005 to replace the EFT system which was public key infrastructure (PKI)-enabled and the settlements were effected on a decentralized mode.

NEFT is more secure, nation-wide retail electronic payment system to facilitate funds transfer by bank customers, between the networked bank branches in the country. This has facilitated the availability of electronic payment modes at more centres. There are 12 daily settlements during weekdays.

The banks are to credit the accounts of the customers for the first four settlements on the same day and for the fifth and sixth settlements, the customers' accounts are to be credited not later than T+1 (next working day).

As on January 7, 2017, NEFT was available at 1,34,363 branches all over the country with 178 participating banks.

In order to popularise the e-payments in the country, the RBI, on its part, waived the service charges to be levied on the member banks, till March 31, 2009, in respect of the real time gross settlement (RTGS) and NEFT transactions. The RBI also provides, free of charge, intra-day liquidity to the banks for the RTGS transactions.

The service charges to be levied by banks from their customers for RTGS and NEFT have, however, been deregulated and left to the discretion of the individual banks. While some of the banks have rationalised their service charges and a few have made it even cost-free to the customers, there are also certain banks that have fixed unreasonably high service charges for providing these services to their customers—even though the RBI provides these services to the banks free of charge.

1.6 Digital Wallet

Digital wallet refers to an electronic device that allows an individual to make electronic transactions. This can include purchasing items online with a computer or using a smartphone to purchase something at a store. An individual's bank account can also be linked to the digital wallet. He may also have his driver's license, health card, loyalty card(s) and other ID documents stored on the phone. The credentials can be passed to a merchant's terminal wirelessly via near field communication (NFC). Increasingly, digital wallets are being made not just for basic financial transactions but to also authenticate the holder's credentials. For example, a digital wallet could potentially verify the age of the buyer to the store while purchasing alcohol. The system has already gained popularity in Japan, where digital wallets are known as "wallet mobiles".

A digital wallet has both a software and information component. The software provides security and encryption for the personal information and for the actual transaction. Typically, digital wallets are stored on the client side and are easily self-maintained and fully compatible with most e-commerce websites. A server-side digital wallet, also known as a thin wallet, is one that an organization creates for and about you and maintains on its servers. Server-side digital wallets are gaining popularity among major retailers due to the security, efficiency, and added utility it provides to the end-user, which increases their satisfaction of their overall purchase. The information component is basically a database of user-input information. This information consists of your shipping address, billing address, payment methods (including credit card numbers, expiry dates, and security numbers), and other information.

Digital wallet systems enable the widespread use of digital wallet transactions among various retail vendors in the form of mobile payments systems and digital wallet applications.

A client-side digital wallet requires minimal set-up and is relatively easy to use. Once the software is installed, the user begins by entering all the pertinent information. The digital wallet is now set up. At the purchase or check-out page of an e-commerce site, the digital wallet software has the ability to automatically enter the user information in the online form. By default, most digital wallets prompt when the software recognizes a form in which it can fill out; if one chooses to fill out the form automatically, the user will be prompted for a password. This keeps unauthorized users away from viewing personal information stored on a particular computer.

1.7 Unified Payment Interface (UPI) of India

UPI is a payment system which facilitates the fund transfer between two bank accounts. This payment system works on the mobile platform. Sending money through the UPI app is as easy as sending a message. The customer is not required to give bank account details for the funds transfer through the UPI payment system.

1.7.1 Benefits of UPI: These are as under:

1. UPI transfers the fund immediately. No restriction of holiday or working hours. The bank strike will also not affect the UPI payments.
2. Customers do not require the bank account number and IFSC code of the recipient.

3. Customer can transact from many bank accounts through a single UPI app.
4. Customer is not required to wait up to 24 hours to send money to a new payee. Anyone would get money immediately.
5. Customer can send bills and get money once the client approves it.
6. Customer can use the cash on delivery without paying cash to the delivery boy. Just approve the bill and the delivery boy would get confirmation.

1.8 RuPay

RuPay is an Indian domestic card scheme conceived and launched by the National Payments Corporation of India (NPCI). It was created to fulfil the Reserve Bank of India's desire to have a domestic, open loop, and multilateral system of payments in India. In India, 90 percent of credit card transactions and almost all debit card transactions are domestic. However, the cost of transactions was high due to monopoly of foreign gateways like Visa and MasterCard. In recent years, the usage of credit and debit cards (called plastic money) has increased manifold. It was thought that if this process of transactions is made India-centric, cost can come down drastically. RuPay facilitates electronic payment at all Indian banks and financial institutions, and competes with MasterCard and Visa in India. NPCI maintains ties with Discover Financial to enable the card scheme to gain international acceptance.

RuPay is a portmanteau of the words *ru*pee and *pay*ment. The colours used in the logo are an allusion to the tricolour national flag. RuPay card was launched on March 26, 2012. NPCI entered into a strategic partnership with Discover Financial Services (DFS) for RuPay Card, enabling the acceptance of RuPay Global Cards on Discover's global payment network outside of India.

RuPay cards are accepted at all ATMs across India under National Financial Switch, and under the NPCI's agreement with DFS, RuPay cards are accepted on the international Discover network. According to the data published by National Payments Corporation of India, there are around 145,270 ATMs and more than 875,000 point of sale (PoS) terminals in India under the RuPay platform. In addition to the ATMs and PoS terminals, RuPay cards are accepted online on 10,000 e-commerce websites.

RuPay cards are accepted at all PoS terminals in India. To enable this, RuPay has certified 29 major banks in India to accept the RuPay card at their respective PoS terminals located at different merchant locations.

1.9 Digital India (DI): Vision 2019

Digital India is an initiative of the Government of India to integrate Government Ministries/Departments with the people of India. It aims at ensuring the availability of government services to citizens electronically by reducing paperwork. The initiative also includes plans to connect rural areas with high-speed internet networks. DI has three core components. These include creation of digital infrastructure, delivering services digitally and digital literacy.

The project, being handled by the Department of Electronics and Information Technology (DeITY) of the Ministry of Communications and Information Technology, is slated for completion by 2019. A two-way platform will be created where both the service providers and the consumers stand to benefit. The scheme will be monitored and controlled by the *Digital India Advisory Group* which will be chaired by the Ministry of Communications and Information Technology. It is an inter-ministerial initiative where all Ministries and Departments will offer their own services to the public such as healthcare, education, judicial services etc. The public-private-partnership (PPP) model will be adopted selectively. In addition, there are plans to restructure the National Informatics Centre (NIC). This project is one among the top priority projects of Prime Minister Narendra Modi.

The Government of India entity Bharat Broadband Network

Limited (BBNL) which executes the National Optical Fibre Network project will be the custodian of DI project. BBNL has ordered United Telecoms Limited to connect 2,50,000 villages through GPON to ensure FTTH based broadband. This will provide the first basic set up to implement DI project and is expected to be completed by 2017.

1.9.1 Digital India (DI): A Critique: DI initiative is a promising initiative of the Indian Government. Many companies have shown their interest in this project. However, it is not free from challenges and legal hurdles. Some believe that DI cannot be successful till mandatory e-governance services are introduced in India. Incomplete implementation of the National e-Governance Plan of India will also adversely affect the success of the DI project. India has poor regulations in the field of privacy protection, data protection, cyber laws, telegraph, e-governance, e-commerce etc. Further, many legal experts believe that e-governance and DI without cyber security are useless. The cyber security trends in India have exposed the vulnerability of Indian cyberspace. Even the National Cyber Security Policy, 2013 has not been implemented till now. In these circumstances, critical infrastructure protection would be a tough task to manage for the Indian Government. The project also lacks the concept of proper e-waste management.

DI initiatives would be required to comply with the civil liberties requirements in general and civil liberties protection in cyberspace in particular. India has not given any importance to privacy and privacy laws so far. Indian Government indulges into mass surveillance in India and projects like Aadhaar and Central Monitoring System are operating without any law and parliamentary oversight. The intelligence agencies of India like Intelligence Bureau (IB) and law enforcement agencies like Central Bureau of Investigation (CBI) are operating for decades without any law and parliamentary scrutiny. DI would further strengthen the mass surveillance activities of the Indian Government if proper procedural safeguards are not implemented and practiced.

The initiative is commendable and deserves full support of all stakeholders. Nevertheless, the initiative lacks some crucial components including inadequate legal framework, absence of privacy and data protection laws, possibilities of civil liberties abuse, lack of parliamentary oversight for e-surveillance, lack of intelligence-related reforms, insecure Indian cyberspace etc. These issues have to be addressed for successful planning and implementation of the DI initiative. Still, DI project is worth exploring and implementation despite its shortcomings that can be rectified before/during its implementation.

If correctly implemented, Digital India project can change the way public services would be delivered in India in the near future.

2

E-governance Initiatives

Information technology has revolutionized lives. Uses of internet and computers are growing immensely. Internet has enabled quick dissemination of information to the public, making the processes all the more transparent, and efficient. The role of the government is to formulate policies and implement them effectively. This involves communication among and within the government departments—upward, downward, horizontal or vertical and with the citizens. Electronic media has immensely eased communication and has improved effectiveness.

2.1 Governance Defined

Governance is a very general concept that can refer to all manner of entities. Equally, this generality means that governance is often defined more narrowly to refer to a particular *level* of governance associated with a type of organization (including public governance, global governance, non-profit governance, corporate governance, and project governance), a particular *field* of governance associated with a type of activity or outcome (including environmental governance, internet governance, and information technology governance), or a particular *model* of governance, often derived as an empirical or normative theory (including regulatory governance, participatory governance, multi-level governance, meta-governance, and collaborative governance). Governance can be used not only to describe these diverse topics but also to define normative or practical agendas for them.

Normative concepts of fair governance or good governance are common among political, public sector, voluntary, and private sector organizations.

The World Bank defines governance as "the manner in which

power is exercised in the management of a country's economic and social resources for development". The Worldwide Governance Indicators project of the World Bank defines governance as "the traditions and institutions by which authority in a country is exercised".

2.2 What is E-governance?

When electronic technology is used for the delivery of government services and dissemination of information to the citizens, the process is called e-governance. In other words, it is the process of policy formulation and implementation through electronic media by the government and its agencies. It is the essence of democracy to bridge the gap between the government and the citizens, making the administrative system more transparent and effective to address to the needs of the citizens. E-governance services can be broadly classified into the following four categories:

1. **Government to Citizens (G2C) Services:** These include services like passport, land records, Aadhar registration, driving license etc.
2. **Government to Business (G2B) Services**: Taxes payment online, e-customs, paperless trade licenses are some examples under this category.
3. **Government to Employees (G2E) Services:** E-office, payroll, e-learning can be included under this category.
4. **Government to Government (G2G) Services:** These services are typically intra-government services, which connect one government department to others. Examples of such services are e-migrate, SSDG etc.

E-governance is an evolutionary phenomenon. In India, the main thrust for e-governance was provided by the launching of NICNET in 1987, the national satellite-based computer network which was followed by the launch of the District Information System of the National Informatics Centre (DISNIC) programme to computerize all district offices in the country for which free hardware and software was offered to the State Governments. In the ensuing years—with ongoing computerization, tele-connectivity and internet connectivity established—a large number of e-governance initiatives, both at the Union and State levels, were taken. The formulation of National e-Governance Plan (NeGP) by the Department of Electronics and Information Technology (DeITY) and Department of Administrative Reforms and Public Grievances (DAR&PG) in 2006 boosted the e-governance process.

E-governance in India is graduating from pilot e-governance projects to bigger Mission Mode Projects. E-governance services thus reckon on the quality of internet service for its stakeholders to reap maximum benefits out of it.

2.3 National e-Governance plan (NeGP)

It aims to "make all government services accessible to the common man in his locality through common service delivery outlets and ensure efficiency, transparency and reliability of such services at affordable costs to realize the basic needs of a common man".

NeGP has a three-tier architecture.

Tier 1 includes common service centres (CSCs) which are the front end delivery points for a range of G2C services. These centres provide transparent services at convenient locations, at affordable costs.

Tier 2 includes state wide area networks (SWANs) and the state data centres (SDCs), both of which provide common and support infrastructure, to allow information to be shared electronically between different agencies of the Government and with citizens.

Tier 3 comprises mission mode projects (MMPs) which transform high priority citizen services from their current manual delivery to electronic delivery. MMP is an individual project within the NeGP that focuses on one aspect of electronic governance such as banking, land records, commercial taxes, etc. NeGP comprises of 31 MMPs which are classified as Central, State and integrated projects. They have clearly defined objectives, scopes, and implementation timelines and milestones, as well as measurable outcomes and service levels.

The use of e-governance services is increasing year-by-year. This is substantiated by data from e-taal, which is a web portal for dissemination of e-transactions statistics of National and State level e-Governance projects including Mission Mode Projects. An e-transaction alludes to a transaction delivering public service using ICT tools to improve access, enhance transparency and reduce response time while also satisfying all of the following four conditions:

1. Service is requested through electronic means (self-access or assisted access) including mobile devices.
2. Workflow/approval process is electronic.
3. Database is electronic/digitized.
4. Service delivery is electronic across the counter services.

A large number of services are being offered by various Ministries/Departments/Organizations at Central and State level. However, the name of a particular service may vary across the states.

2.4 E-governance and Internet Infrastructure

Internet is the backbone of e-governance services. In order to usher e-governance services, National Informatics Centre (NIC) has been involved in supporting the government in improving the internet/intranet infrastructure for enhancing the provisioning of e-governance services in the country.

A country-wide satellite-based communication network NICNET provides connectivity to Central Government Ministries/Departments, States/UTs and districts of India. In a number of States, State Wide Area Networks (SWANs) are also being integrated with NICNET.

The Government approved the scheme for establishing SWAN across the country in March 2005. Under this Scheme, technical and financial assistance is provided to the States/UTs for establishing SWANs to connect all State/UT headquarters up to the block level via district/sub-divisional headquarters, in a vertical hierarchical structure with a minimum bandwidth capacity of 2 mbps per link. Each of the State/UT can enhance the bandwidth up to 34 mbps between SHQ and DHQ and up to 8 mbps between DHQ and BHQ depending upon the utilization.

To monitor the performance of SWANs, the department has mandated positioning of third party auditors (TPAs) in the States/UTs.

Ministry of Electronics and Information Technology (MeitY) proposed that high speed broadband connectivity would be made available to all 2.5 lakh gram panchayats in the country under BharatNet programme (which is presently being implemented by the Department of Telecommunications). Robust connectivity would enable Central and State Governments to deliver various G2C, G2B and G2G services efficiently, while also enhancing delivery of other social sector services including e-health, e-education, e-agriculture etc.

To develop rural wi-fi network infrastructure and in an attempt to provide means to transform Indian villages into *smart villages*, CSC e-Governance Services India Limited has taken an initiative called *e-Choupal* (w-ifi choupal).

A number of services are today provided by the Central and State Governments. These include ration cards, caste certificates, income certificates, certificates for proof of residence, passports and similar other services. It is important that these are available within a prescribed time line. Failures and deficiencies in delivery of public services lead to dissatisfaction and public anger against the government. Various government agencies have taken initiatives to notify services and time-limits within which these are to be availed of. An important aid to delivery of services can be the use of e-governance and technology.

The use of technology in delivery of public services will need to be expanded rapidly to reduce delays and discretion used to the advantage of the citizens.

2.5 Unique Identification (UID) System

In a citizen-centric system of governance, the citizen's satisfaction becomes the measure of success of both service delivery as well as programme delivery. A citizen's identity, therefore, becomes important both in developmental as well as regulatory administration. The absence of a reliable system for such purposes has been an impediment to improving targeting of developmental schemes and reducing leakages in the delivery system.

Many major individual-oriented Government programmes incorporate a provision for collection of information at the individual/family level. In most cases, this is undertaken as a de novo exercise without reference to similar exercises undertaken in the past by other government departments and sometimes even by the same department. The absence of a system of updation of such purpose-specific databases and the lack of a system for corroboration among such departments are also factors leading inevitably to expensive, time consuming and error-prone de novo surveys for data collection for each scheme. To create a common platform for service/programme delivery, UID system has been introduced in India.

Development of unique ID numbers with biometrics to establish proof of identity is an important area which has been thrown open for use of technology. This would help in providing services to various users and result in controlling fake cards and thereby bring enormous savings. The financial inclusion services are also feasible by using UID and the telecom services. It is also possible to use technology for expanding banking correspondence and thereby expand the extent of banking services.

UID System is envisioned as a means for residents to easily establish their identity, anywhere in the country. It is an important step towards ensuring that residents in India can access the resources and benefits they are entitled to. A resident can enrol for a UID number by providing basic demographic as well as biometric details to the enrolling agency. The enrolling agency will transmit these details to a central UID server. The server will then perform a de-duplication check using the resident's key demographic and biometric fields against existing UID records in the database, to ensure that he/she does not already have a UID number.

Once the check confirms that a duplicate record does not exist, the central system will issue a UID number to the resident. The resident can then use the number with different service providers, who can verify his or her identity online. The agency has to transmit the UID number and information provided by the resident to the UID server, and the server immediately responds with a yes or a no.

UID can have a significant impact on service delivery. The existing patchwork of multiple agency databases in India gives individuals the incentive to provide different personal information to different agencies and also impersonate someone else. In the UID infrastructure, all resident records are stored in a central database, and each new entry is de-duplicated. Consequently, residents can only have one UID number, which is mobile and can be used anywhere in the country. The lack of duplicates, and accuracy and mobility in identity verification, reduces opportunities for fraud and enables agencies across the country to provide residents with targeted, effective services and benefits.

2.5.1 Smart Cards: The UID Project will eventually become the underpinning of the Citizens Smart Card Project of the Ministry of Home Affairs—both are aimed at identifying citizens, providing them a unique identity and in case of the MHA project, also will provide a Smart Card which has all this information.

The smart card would have a memory partitioned into distinct modules representing different entitlement groups for which free services or implicit/explicit subsidies are given. These include food and nutrition, energy (kerosene, LPG, electricity), education services, health services, civic amenities and services (drinking water, latrines/sanitation), employment and farming (fertilizer, irrigation etc.).

Any subsidy received by any individual would be entered on his/her smart card when the goods or service is delivered/charged for by the authorized supplier (for example, the fair price shop, kerosene/LPG dealer, fertilizer outlet). The rules and regulations for delivery of subsidy and its reimbursement to the goods/service supplier would be defined by the concerned department.

The data entered on the smart card would be accessible by all monitoring/evaluating agencies so that they can put together a picture of what subsidies are being received by whom, as well as those who are not receiving a subsidy for which they are eligible.

2.6 Electronic Technology in Tax Administration

2.6.1 E-filing of Returns: E-filing of income tax return is the process of electronically filing returns through internet which can be filed at any time at any place. While e-filing of income tax return is mandatory for a company and a firm liable to audit under Section 44AB, it is optional for other assessees. Similarly, Government has introduced e-filing of returns of tax deducted at source. It is mandatory for corporate deductors to furnish their TDS return in electronic form with effect from June 1, 2003. Further, it has been made mandatory for Government deductors and firms liable to audit under Section 44AB with effect from assessment year 2004-05. Deductors have to file e-TDS returns quarterly since assessment year 2005-06.

National Securities Depository Ltd. (NSDL) has been appointed as the e-TDS intermediary by the Income Tax Department. NSDL receives e-TDS returns from deductors on behalf of the Income Tax Department. Deductors can submit e-TDS returns through TIN facilitation centres established by NSDL or directly upload through NSDL website.

2.6.2 Online Tax Accounting System (OLTAS): Income Tax Department operationalised OLTAS in July 2004. The new single copy challans have been introduced with effect from July 2005. The collecting bank branch will put a rubber stamp on the challan and its counter foil indicating a unique challan identification number (CIN), BSR code and challan serial number. The collecting bank has to capture the entire data of the challan and transmit it electronically to the Income Tax Department. The information received from banks is used by the Department to give credit for the tax paid based on CIN. This is a very significant step for creating tax information network.

To sum up, governance can be much more interactive if extensive use is made of all channels of communication including print and electronic media, social media, electronic boards in public places, written materials, website, and other methods that have a wide reach and are able to convey messaging directly and swiftly.

India recognizes IT as the fastest and the most advanced vehicle of change for all-round progress and development of the country. In view of the potential of IT, the Government has advocated widespread proliferation of IT in the country and adopted various policy supports for promotion of IT in the fields of e-governance, empowerment of the people, education, industry, health, rural development, agriculture, and tourism.

2.6.3 Permanent Account Number (PAN): PAN issued by Income Tax Department is the critical element in capturing income and expenditure of a person. The Department identifies the assesses/persons with PAN which is a unique 10 digit alpha-numeric number. Obtaining PAN is compulsory not only for income tax purposes but also for certain other purposes/transactions. PAN is to be obtained only once for ever. An assessee need not obtain a new PAN, even if he is transferred to any other place.

2.6.4 Digital Signature: A person desiring to furnish his return of income electronically may sign it digitally or manually. For signing the return digitally, he is required to obtain a digital signature. A digital signature is the electronic signature issued by the certifying authority that shows the authenticity of the person signing the same.

2.6.5 Goods and Service Tax Network (GSTN): Among the various steps that have been taken for the introduction of goods and services tax (GST) is the establishment of a strong information technology (IT) infrastructure. For this purpose, the Government has set up an Empowered Group under the chairmanship of Nandan Nilekani. Significant progress has been made in the conceptualization and design of the GSTN—a common portal for the Centre and States that will enable electronic processing of the key business processes of registration, returns, and payments. This portal has started operating after the rollout of GST in 2017.

2.7 E-commerce

E-commerce involves individuals and business organizations exchanging business information and instructions over electronic media using computers, telephones and other telecommunication equipments. Such form of doing business has been in existence ever since electronic mode of data/information exchange was developed, but its scope was limited only as a medium of exchange of information between entities with a pre-established contractual relationship. However, internet has changed the approach to e-commerce; it is no longer the same business with an additional channel for information exchange, but one with new strategy and models.

2.8 Electronic Services for Rural Development

2.8.1 E-choupal: E-choupal offers farmers the information, products and services they need to enhance productivity, improve farm-gate price realisation, and cut transaction costs. Farmers can access the latest local and global information on weather, scientific farming practices, as well as market prices at the village itself through this web portal—all in Hindi. E-choupal also facilitates the supply of high quality farm inputs as well as the purchase of commodities at the farm.

Given the literacy and infrastructure constraints at the village level, this model is designed to provide physical service support through a *choupal sanchalak*, himself a lead farmer, who acts as an interface between the system and the farmers.

The contents of this site in their entirety are made available only to the registered *sanchalaks*.

2.8.2 E-panchayat: E-panchayat is one of the Mission Mode Project (MMP), currently being implemented with a vision to empower and transform rural India. As a first step towards formulating the project, the Ministry of Panchayati Raj constituted an Expert Group in June 2007 under the Chairmanship of Dr. B.K. Gairola, Director General, NIC, Government of India. The Expert Group was entrusted with the task of assessing the IT Programmes of the Ministry of Panchayati Raj and recommending cost effective solutions along with the cost implications. Adopting a consultative approach, the Committee interacted with States/UTs to assess the existing status of computerization up to the gram panchayat level, including the initiatives undertaken by the State Governments. The recommendations of the Committee formed the basis for the conceptualization of e-panchayat.

The e-panchayat project holds great promise for the rural masses as it aims to transform the panchayati raj institutions (PRIs) into symbols of modernity, transparency and efficiency. This is a one of its kind nationwide IT initiative introduced by Ministry of Panchayati Raj that endeavours to ensure people's participation in programme decision-making, implementation and delivery. The project aims to automate the functioning of the 2.45 lakh panchayats in the country. The project addresses all aspects of panchayats' functioning including planning, monitoring, implementation, budgeting, accounting, social audit and delivery of citizen services like issue of certificates, licenses etc.

Subsequently, e-panchayat scheme was subsumed by Rajiv Gandhi Panchayat Sashaktikaran Abhiyan (RGPSA).

3

Importance, Uses and Applications of Internet

Internet is a revolutionary technology that has transformed the lives of people in an unimaginable way. As one hears the phrase *industrial revolution*, one can recollect the era in late 18th and 19th centuries, when rapid changes were taking place all over the world, led by England and America. Those were the changes which transformed the way manufacturing was done. This generation is witnessing another such evolution in the field of technology. Modernization has made machinery an indispensable part of human life, without which one cannot even think of travelling from one place to another. The proliferation of internet is being seen as another such type of industrial revolution which makes it imperative to assess the validity of this belief.

3.1 Growth of Internet: India and the World

Today, internet has become the lifeline of modern day businesses and helps common people in accomplishing their day-to-day tasks. World economies have progressed in the field of security and surveillance and common people rely on it to meet their information and education needs. This is just the basic of internet. Its depth and intensity to influence human life is way beyond what any of us could ever imagine. Today, there are 7 billion people dwelling on this blue planet of ours. We live in a society where everyone wants to stay connected and have their needs fulfilled in the shortest possible time. Now, let us look at some numbers to have a clear picture of how internet has become the most important innovation of this century and how it is impacting our society.

There are 3.2 billion internet users in the world, 207 billion mails are sent everyday from internet, 3.8 billion

YouTube videos are watched every day, 803 million tweets are sent in a day and 36 million is the number of Amazon purchases made in a day (ICRIER, 2016). And, according to Hostingfacts.com, 2016, there are 1.5 billion active Facebook users in the world, 2.9 billion Google searches are made every day, 2.7 million blog posts are published every day, and 966 million is the number of websites in the world. Another chilling fact is that only 44 percent of the web traffic is from humans; a massive 56 percent of web traffic is from bots, impersonators, hacking tools, scrapers and spammers.

The above facts provide enough evidence of our dependency on the internet in modern era. With time, this dependence is going to increase even more. About half of the world population is yet to be connected to the internet. The story is progressing almost on the same lines in India as well.

According to a report by Statista, India is the second largest online market, with over 460 million internet users. However, it must be noted that only internet user figures give a partial picture as only 26 percent of the Indian population accessed the internet in 2015, despite the vast user base. The study also found out that men dominated the internet usage in India with the percentage of men and women internet users being 71 percent and 29 percent respectively. There was another interesting finding that the majority of the population using internet accessed it through their mobile handsets in 2016, the figure being 323 million users. The figure is expected to increase further in the years to come, the projections amounting to 524.5 million in 2021. As of 2016, the estimated internet users in India stood at 262 million in urban areas and 109 million in rural areas (Statista, 2016).

As per Internet live stats data (Internet live stats, 2016), the share of India's internet users in the world digital population is 13.5 percent, with the penetration rate in India being 34.8 percent. The penetration of internet in India is increasing. The percentage of population using internet was only 27 percent in 2015 and 18 percent in 2014. Internet user base has increased by 30.5 percent in 2016.

Statscounter, an independent analytics company, found that internet usage worldwide on mobile and tablet exceeded that on desktop for the first time in October 2016. As measured by the company, mobile internet usage was 51.3 percent while desktop internet usage was 48.7 percent. The data was based on over 15 billion page views per month to over 2.5 million websites. According to the report, 75 percent of India's internet usage occurred through mobile phones. It is interesting to note that while emerging markets have shown a growth in internet usage on mobile and tablets, mature markets like UK and US, continue to have higher internet usage on desktops (StatCounter GlobalStats, 2016).

However, this growth has not been inclusive so far. 850 million Indians still do not have an internet connection. Every day, almost 70,000 people in India use internet for the first time (India Telecom 2016; Transforming India, 2016).

3.2 Motivation and Barriers to Internet Usage

The rate of internet adoption is increasing at a fast pace worldwide, but there are still millions left out. Governments have recognized its importance and therefore are making efforts to increase the level of internet penetration in their respective countries. In India, about 34 percent of the population is using internet, internet penetration being 18 percent in rural India. As more and more people have started using internet, there are many who are yet to see how their lives can be changed with internet. Those who use internet have some motivation to use it and those who do not are trying to break the barriers that restrict their internet usage.

According to TRAI, only 31.35 crore people, out of a total wireless subscriber base of 103.37 crore, use internet on their handsets. Also, the growth observed in wireless subscriber base which used internet was a mere 16 percent in 2015-16. The question that arises here is what are the reasons behind such a meagre growth for a country having the world's second largest population? According to TRAI, following reasons seem to be plaguing the adoption of wireless internet in the country:

1. Low levels of literacy and digital literacy.
2. Low perceived utility of the internet.
3. Lack of adequate internet coverage.
4. Non-affordability of data among poor customers.

According to Katz and Aspden (1997), the main motivation/reason for people to use internet is to communicate with others and get information of general interest to keep up-to-date. Commercial reasons and contacting new people appeared to be minor reasons for using the internet. However, as this study was done in 1997, the dynamics ruling these motivations have changed pretty much. In 2016, major businesses have their presence online. Even brick-and-mortar businesses made use of internet in conducting their day-to-day operations. Internet makes us meet new people more conveniently than ever before. That is why we are witnessing a spurt in number of social media and dating sites.

Another point made in the study by Katz and Aspden (1997) is that *social-personal development* is the key driver of internet usage among current and former users. Non-users have emerged to be more strongly drawn to the internet for business reasons and the opportunities for staying up-to-date. The study also found that long-time users were more likely to use online services as compared to recent users. It has also outlined some barriers to get started for the non-users such as cost of internet services, lack of comprehension, difficulty to access internet, complexity and uncomfortable sitting in front of a computer.

3.3 Impacts of Internet on Day-to-day Life

Pulitzer award winning writer Thomas Friedman in his book *The World is Flat* has described the flatteners that shaped the 21st century global economy. These flatteners, beginning with the fall of Berlin wall, allowed us to think of the world as a *seamless whole*, to see it as *a single market, a single ecosystem and a single community*. Some of these flatteners have emerged as a result of internet. Internet has now started to have an impact on our daily lives. We resort to internet for most of our needs, be it communication, education, travel or shopping. About 70-80 percent of the internet users use social media. Technologies touching life-travel sites are the most used sites, even more than the shopping sites, followed by banking.

Quettera Software (Chafekar, 2015) analyzed thousands of users in India based on their usage data. It found that daily usage of shopping apps is about 402 times higher than in US. At an app level basis, total time spent on WhatsApp dwarfs all other apps. An average Indian user spends 25 percent of his/her total mobile usage time on WhatsApp messenger. This number is 2.5 times higher than of an average US user.

Internet is undoubtedly one of the biggest inventions of this century. Another such invention which has actually helped in internet proliferation and its best possible use is mobile phone. Smart phones have made it possible to easily access internet as and when required. India too has been reaping its enormous benefits.

3.4 Internet and Economy

DePrince Jr. and Ford (1999) assessed the impact of internet on the structure and growth potential of US economy. The internet economy in the US had been growing at a three digit growth rate annually and contributes significant portion of country's GDP along with the employment of huge number of workers. Although the growth rate of the internet economy will decelerate in coming years, its share in the economy's total output and employment will continue to increase rapidly.

The internet economy comprises of four sectors: (a) hardware providers, (b) software/application providers, (c) intermediary/market makers and (d) companies engaged in e-commerce. The first two help in building and operating the required infrastructure.

In the consumer sector, producers and distributors of high value-to-weight and high value-to-volume products like personal computers and peripheral equipment, computer software, books, video discs are already experiencing rapid growth of internet-driven sales. The transportation industry is also being heavily impacted by the migration toward e-commerce. As consumers are getting home deliveries for their online purchases, shipping companies are experiencing a surge in the volume of their transactions. Health related transactions that can be processed on the internet include medical record keeping, health provider searches and appointments, health/medical information directories, insurance applications and payments and pharmaceutical orders and records, are resulting in major cost savings.

Consumers of products and services use the internet to scan global markets on a real-time basis. The explosive growth in effective search engines has reduced cost of searching and making transactions over the internet. Internet clearly intensifies the price-war and thereby reduces inflationary pressures engendered by segmented and disjointed markets.

Internet has had a profound impact on the services sector. In the airline industry, consumers have shifted from traditional modes of booking to website booking, where they get bonus miles, more discounts and special services. Financial services industry is also undergoing a transformation to use internet as the main communication platform. Many large firms now offer online trading of stocks, bonds and other financial instruments, which bypass the human contact generally involved in such transactions. In addition to marketing services, internet has also facilitated the introduction of a wider range of secondary markets for various real assets to solve their liquidity problems.

The authors also found that there is a transition taking place in the distribution channels (Amazonic and/or Dellphic) which is set to reshape the micro-economic structure and operating characteristics of various industries. In Amazonic distribution, an end user orders products directly from a distributor that maintains a warehouse inventory of products. Those products are ordered and reordered from producers, in response to sales, usually via wholesale-level internet or private intranet transactions. The end user gets the product directly from the distributor. This channel bypasses retailers and thereby undermines the retailers' business.

Conversely, Dellphic distribution involves direct contact between end users and producers in which there is no inventory of finished products anywhere in the distribution channel. In this model, end users respond directly to producers' promotional efforts by ordering custom-made products telephonically or via the web, which are then produced and shipped directly to the buyer. Products and services such as sale of personal computers and books, stock trading, and travel services are already migrating very rapidly towards this transition.

3.5 Leisure and Recreation

3.5.1 Travel and Leisure: With the commercialization of internet, the nature of society has changed in many ways. In recent years, the economic power of internet has grown and further reconfigured the world we live in. Today, internet can be seen as a platform that assists in continuous technological innovations, encourages new business practices and changes the competitive landscape of various industry sectors. With technological innovations, internet has changed from *publishing-browsing-platform* to a *participation-interaction-platform*. It has enabled new business models and fostered changes in social life.

Internet provided the foundation for the expansion of new systems linking consumers and computer reservation system and global distribution system (CRSs/GDSs), which certainly lowered the entry barriers of new players. Thus, in late 1990s, a series of online travel agencies like Expedia, PreviewTravel, Priceline, and TravelBids began to offer direct access to the travel products. The innovations in the travel and hospitality sector have been widely accepted by the consumers.

Researchers have clearly identified internet as a platform that cannot be overlooked or ignored by tourism organizations while forming their marketing mixes owing to its advantages such as global accessibility, convenience in updating, real time information service, interactive communication features and unique customization capabilities. Internet is successful in swift identification of customer needs and establishment of direct contact with consumers, offering them comprehensive, personalized and up-to-date information. Tourism suppliers can remotely control their servers to display information on products/services at an electronic speed. As far as the travellers are concerned, internet allows them to communicate directly with tourism suppliers to seek information. It also empowers them to develop and purchase their own itineraries anytime and anywhere.

3.5.2 Games and Gambling: Internet provides tremendous educational benefits to children and adolescents, such as enhanced information access, better visual intelligence skills, and enhancement of teacher-student communication. But it has some serious implications too. A time has come when parents are literally forcing their children to go outside and do some physical activity because outdoor games have now been replaced by mobile phones and online gaming. A study done in China (Jiang, 2014), examined the interrelationship between internet connectedness, online gaming, internet addiction symptoms and academic performance decrement among the internet dependent young people in China.

In China, internet addiction has been identified to be a growing problem among young people, who account for about 60 percent of internet users, of which 14 percent are adolescents, equally spread in rural and urban areas. It is an alarming situation because excessive internet use can be pathological and addictive, and is treated like an internet addiction disorder having various symptoms that include substance based addictions, like tolerance, withdrawal, preoccupation, numerous unsuccessful attempts to cut down the use of internet, and feeling restless without internet.

Gambling in India is heavily restricted except for selective categories including lotteries and horse racing. The critics of gambling claim that it leads to crime, corruption and money laundering while those in favour of a regulated gambling system in India argue that it can be a huge source of revenue for the state. Whatever may be its harmful implications, gambling has seen a tremendous growth with the advent of ICT.

The fast technological developments in internet are increasing the global potential of gambling. Internet gambling is set to multiply in times to come, primarily because of its easy access and participation. Sophisticated gaming softwares, integrated e-cash systems that include multi-currencies, multilingual websites, increased realism through webcams, live betting mechanisms and improved customer care solutions, all can be attributed to the increased popularity of internet gambling. Internet gambling is global, accessible and has 24-hour availability. It offers convenience gambling. The initial set-up costs are also minimal in comparison to a casino.

Easy access to internet gambling has major social implications. It does not provide for gate-keeping of vulnerable set of people that include adolescents, problem gamblers, drug/alcohol abusers, the learning impaired, etc. It can extend its claws to workplaces as well, impacting work efficiency and employee productivity. Unscrupulous operators indulge in embedding (pop-ups related to gambling websites), circle jerks (never ending loop of similar websites) and online customer tracking (collecting customer data) to lure internet users into internet gambling.

There are some differences between traditional gamblers and internet gamblers as pointed in the study. First, internet gamblers stop gambling when they run out of money they had set aside for this purpose. But traditional gamblers continue gambling even if they have to borrow money to satiate their thirst for it. Second, traditional gamblers enjoyed gambling as a means of escape but internet gamblers find the real gambling environment at their home satisfying enough. Third, internet gambling is highly competitive and more exciting as compared to traditional gambling because of the introduction of person-to-person gambling. Fourth, internet gambling satisfies the psychological need of self-esteem. Fifth, physiological effects are more prominent in traditional gamblers as against internet gamblers.

There is not enough support for the argument that internet gambling can be doubly addictive, majorly because internet and gambling alone are addictive too. But it also cannot be denied that internet has made gambling more convenient, anonymous and enjoyable and caters to the psychological needs of users.

3.5.3 Internet and Pornography: Pornography has become mainstream over the last two decades. Internet has provided easy access to sexually explicit material to people of all ages. But its popularity has been growing among adolescents like never before. This easy access comes at the cost of risks of internet addiction, cyber bullying, sexual victimization or harassment. Internet has also increased the frequency of such incidents where adolescents are exposed to obscene material.

A small, but growing body of research indicates that adolescents are increasingly struggling with compulsive internet use (CIU) and compulsive behaviours resulting from internet pornography and cybersex. Research suggests that there is a link between adolescents exposed to internet pornography and their sexual beliefs. Pornography can change viewer's perceptions about sex which may differ from those instilled by their families and schools. This may lead to dissonance and conflict in sexual beliefs and is attributed to increased sexual uncertainty.

Sexually reactive children and adolescents (SRCA), who are frequently exposed to sexually explicit material, are more likely to engage in "coerced vaginal penetration and forced sexual acts such as oral or digital penetration, to express sexually aggressive remarks (obscenities), and to engage in sex with animals". Also, higher the exposure to sexually explicit material, lower will be the degree of social integration. It also negatively affects their self-control and decision-making abilities. Children and adolescents who themselves are not involved in internet pornography but live amongst people who consume it frequently are also at high risk of consuming such content and indulging in problematic sexual behaviour in future.

4

Internet and Human Capital Formation

4.1 Internet for Educational and Skills Development

4.1.1 Paradigm Shift: Pollock and Wilson (2002) recognize that the use of internet for educational purposes has been growing rapidly as internet assignments and internet-mediated instruction replace face-to-face class time. They add that internet has been helpful in containing the rising costs of higher education, accommodates more diverse body of students and may also be effective than courses that are delivered face-to-face. It further validates that there are three functions of computer enhanced instruction model—information, instruction and communication—that aim at increasing student participation, enhancing written communication skills and focusing the learning experience on interaction and reflection.

Agarwal and Day (1998) reassure that internet has resulted in two major advantages for students. First, it offers a new medium of interaction that complements classroom instruction and facilitates learning. Second, it offers the opportunity to learn, use internet technology and yield positive externalities for future academic and career paths.

According to Hinson (2006), internet can improve the productivity and efficiencies of university academicians as they can use internet to improve their skills and get any knowledge and information.

Fung and Pun (1997) concluded that the main goal of acceptance of rapid developments in technology by the education sector is to support learning and teaching to make the best schools, teachers, and courses available to all students without regard to gender, distance, resources, or disabilities, i.e. teaching without discrimination.

4.1.2 Information Transmission and Retrieval: Agarwal and Day (1998) record that recent developments in technology have made transmission of information possible through computer communication and conferencing that include e-mail, mailing lists, newsgroups, interactive messaging or chat sessions, video conferencing and information access, retrieval and use through FTP, www etc. E-mail, mailing lists, and chat software promote instructor-student interaction, and web page dissemination of information provides easy access to class syllabi, schedules, lecture notes, projects, and assignments.

Hinson (2006) also maintains that internet offers a range of capabilities for retrieving information, providing access to multimedia data and allows users to utilize information at remote locations.

Fung and Pun (1997) acknowledge that wealth of information on internet is growing exponentially each day and has replaced the dominant sources or vehicle of knowledge, i.e. books and papers. Utilizing these traditional ways of obtaining information was a tedious job when it came down to searching for a specific piece of information through the index or table of contents. In addition to this, the information was confined to the pages of a book, which became a handicap, as information must always be updated and available on demand at immediate basis. Internet has not just rendered these sources of information as uneconomical in terms of time, costs and space, but has also equipped users with multimedia and hyper-linking capabilities that allow information to be presented in the form of sounds, videos and pictures with no limit on number of users who could use it simultaneously.

4.1.3 Interaction and Communication: Fung and Pun (1997) observe that internet breaks the boundary of traditional classroom by allowing teachers and pupils to have immediate connection with the outside world.

Agarwal and Day (1998) add that students find value in electronic communication as it provides better access to instructor, more interesting assignments and advocate future use of e-mail in their academic and professional careers. The authors note that greater interaction between the instructors and the students and hands-on learning of new concepts are results of effectiveness of internet usage.

Pollock and Wilson (2002) also confirm that computer-enhanced format of instruction was marked more favourable for helpfulness of discussions with other students, ease-of-contact with the instructor, and opportunities for expressing and sharing ideas. It elaborated that instructors rely on internet to provide course material, direct assignments and readings, and obtain feedback from students, whereas students can read, process, and analyze information, discuss the issues with their peers, and complete assignments.

Hinson (2006) notes that academicians are able to communicate with their worldwide colleagues through internet and discussion groups allow users to follow issues of interest and keep themselves up-to-date on these issues. It also serves as a tool for supporting networking among professionals in different geographical locations.

4.1.4 Attitude towards the Subject: Pollock and Wilson (2002) maintain that the students who enrolled for computer-enhanced format, expressed greater willingness to take up other courses that employ similar formats and had higher level of overall satisfaction.

Agarwal and Day (1998) support this view and confirmed that computer conferencing allows for increased interaction and discussion, which results in increased ability of students to apply theory and thus stimulate their perceptions and interests towards their subjects. The authors observe that the internet-enhanced group expressed a significantly higher likelihood of attending a lecture given by an economist and was more likely to consider economics as their favourite subject, used economic concepts to analyze situations more frequently, and disagreed about finding economics dull.

4.1.5 Collaborative Teaching: Fung and Pun (1997) opine that nothing in history has been as powerful as the internet in terms of changing traditional roles of teacher as an expert and students as passive receptacles for handover of knowledge. Active and expression learning has shifted traditional information-giving to pupil's participation in information-gathering, interpretation and use. Internet is like a vehicle for teachers to create high performance learning environments through which the goals of various dimensions of schooling can be accomplished. Through resources available via the internet, students may be provided opportunities to engage in authentic, challenging activities; teachers can set meaningful, challenging activities as the centre of instruction; and classroom approaches that support such a paradigm might include collaborative learning, heterogeneous groupings, teachers as facilitators, performance-based assessment, peer-to-peer mentoring, multidisciplinary curriculum, interactive modes of instruction, student exploration, and extended blocks of time.

Agarwal and Day (1998) add that the point of using internet is to add value to classes and to meet the challenges of teaching, as it substantially reduces the contact time through email and discussion lists that allows students to observe the real-life implications of the theory learnt in class. Thereby, the authors conclude that internet serves as an information revolution and its use in pedagogy is beneficial whenever interaction, discussion, research or transmission of information is involved.

4.1.6 Enhanced Learning: Pollock and Wilson (2002) observe that students who studied under computer-enhanced instruction model exhibited an edge in creative thinking, further facilitating creative application of assignments to course themes and contents. Furthermore, computer-based learning models required higher level of interaction between the students and the material and thus fosters more thorough and meaningful learning than standard lecture delivery.

Agarwal and Day (1998) support this opinion by adding that computer communication and ease of information retrieval through the internet allows higher development of critical thinking and problem solving, foster independence and autonomy, and permit greater interaction. Moreover, better correspondence fosters thought and interest in the subject matter and hands-on experience which makes understanding and learning process more active.

As per Fung and Pun (1997), students can learn by doing in a more authentic sense, using real life situations and simulations at times and they may learn not only from their teachers but also from many others by becoming both publishers and browsers on the internet. Internet will not just change classroom processes, but also pupil's learning through an ultimate home infrastructure as web technology has opened new channels for direct sales to customers. An example is the internet for learning pages that provide learning materials to pupils and students directly. The spectrum of their subject contents, design and attractiveness to pupils is beyond the competition of the curriculum in any single school.

4.1.7 Students' Empowerment: Fung and Pun (1997) acknowledge that web technology has shifted the ability to control information flow from the information provider to the users. However, the authors could not rule out the possibility of this empowerment being a threat to many teachers. Moreover, pupils can now acquire their own communication links for exploring topics beyond the boundaries of the traditional concept of school and they can quickly access learning resources at any hour of the day while studying at home, or in a library or any place with internet connection.

4.1.8 Improved Quality of Education: Agarwal and Day (1998) suggest that internet enhancement of courses facilitates communication between the instructors and students for easy access of information that has promoted the use of economic data and real-world applications to enhance the teaching of the theory. This provides real increase in quality of education. Besides this, higher simulation of interest and better use of class time and communication of ideas was noticed to be present in the internet group and thereby resulted in better overall assessment of the instructor and his feedback and interest in students. As a result, instructor effectiveness improved due to facilitation of learning and communication of ideas and information.

The findings of Pollock and Wilson (2002) also showed that students in the computer format showed double increase in their knowledge improvement as compared to that of students in traditional lecture formats.

4.1.9 Contradictory Concerns: Pollock and Wilson (2002) raised some concerns that challenge each other. First, internet has allowed some people to work at their own pace, but they also feel that it has reduced face-to-face time with instructors. Second, some people found it interesting to use the internet for educational purposes, whereas some had to face network problems. Third, forum discussions yielded better results for some, while others went confused in figuring the flow of the conversation. Fourth, instructor's enthusiasm and knowledge can provoke interest for some but others had to wait for days to get responses from instructors. Fifth, internet can be interesting for some but others may find it hard to understand, as it might not explain well. Sixth, altogether, internet can be an enjoyment and fun for some, while others might think of it as boring and repetitive.

4.1.10 Negative Aspects: According to Fung and Pun (1997), the entire schooling system will be re-engineered in the era of internet, where students are more detached from a single school and the teachers therein and open to more opportunities of interacting directly with other schools and *knowledge providers* on the internet. On top of this, web technology has made both publishing and retrieval of information accessible to many, due of which schools will now have to be gatekeepers in short-listing books and educational websites that students can read and visit as finding relevant information becomes difficult in such a loosely controlled and independent environment. Internet is anarchic in a way that it is accessible by anyone anywhere in the world and anyone anywhere can publish anything. Without proper management, the internet could become a potential threat to the well-being of a school and its pupils when they are internet connected.

4.2 Internet for Health and Medical Care

London and Dash (2016) brief that many healthcare processes can be digitized, including appointments, logistics, patient records, admissions, human resources and rotation management, and billing.

4.2.1 Advanced Analytics: London and Dash (2016) argue that healthcare industry must harness the power of data through advanced analytics and transparency as available applications that rely on healthcare data analytics support patients in understanding and managing their medical condition, influencing their medical care and help in identifying the drug to which they are likely to respond best. It elaborates that the process of making crucial clinical decisions about diagnosis and treatment will increasingly be supported by algorithmic tools that compare a patient's clinical and other data with large datasets and draw on the full body of scientific literature.

4.2.2 Reduced Costs: London and Dash (2016) propose that healthcare costs can be reduced up to 10 percent as the collection and publishing of data on outcomes and quality of care can also allow systems to modulate tariffs and orchestrate competition among providers based on their quality of care and will be important for raising the overall standard of care across the healthcare systems. Collaboration of information and care profiles will lead to improvement of effectiveness and efficiency of maintaining patients at home, thereby avoiding unnecessary hospital stays, improving clinical outcomes, and reducing the costs of care.

4.2.3 Improved Quality: London and Dash (2016) assert that in addition to providing efficiency gains, automation can improve patient care like remote monitoring of intensive care units via patient sensors and a central control room can lead to reduction in mortality rates and the average length of hospital stays. It adds that digitization can also bring significant benefits in the area of clinical trials, such as improving the efficiency, reliability of clinical data collection, trial monitoring and optimizing trial design through the use of modelling tools. Moreover, enhancing clinical decision-making and the monitoring of treatment protocols can significantly reduce variability and increase adherence to good clinical practice, thereby promoting healthier behaviours.

4.2.4 Patient Engagement: Hospitals use portals to drive patient engagement. Internet Health Management (2016) defines patient engagement as greater involvement of patients in healthcare that results in their good health and ensures its maintenance. Many hospitals in Western countries have taken some form of patient engagement initiatives that are information technology driven and help in delivering better healthcare services. Most popular tools used by hospitals to drive patient engagement are web portals, secure emailing, electronic health record, online scheduling and reminder tool and social media.

4.2.5 Repercussions of Internet Addiction: Young (1998) notifies that the initial symptoms of internet addiction are depression, bipolar disorder or anxiety, which later grow into compulsive use of internet that is characterized by preoccupation with the internet, lying about the behaviour, a loss of interest in other activities, frequent time distortion, sleep deprivation, increased social isolation, being secretive about online activities or a sudden demand for privacy.

Internet addiction may increase risk of depression and anxiety. Excessive use of internet may be responsible for increasing the risk of depression, anxiety and other mental problems, especially among college going students. Those having internet addiction face more trouble in accomplishing day-to-day tasks and have attention deficiency that leads to poor planning and time management skills.

4.2.6 Bone Health: Continuous texting on social media/mobiles and gaming through the day can cause injuries arising out of prolonged stress positions and repetitive motions. Leading orthopaedic experts have warned about wrist and finger joint pain, repetitive stress injuries to the hands, arms and back, wrong postures, physical stress and problems in cervical and spinal cord. Neglect of such conditions can further result into arthritis of fingers, blackberry thumb, carpet tunnel syndrome and disturbed spine.

4.3 Influence of Internet on Habits

4.3.1 Interpersonal Relationships: Mokhtari, Reichard, and Gardner (2009) note that internet influence on social life appears to revolve around whether the internet makes things more efficient for people by freeing up additional time for socializing and other activities or whether internet use is a potentially isolating activity that competes with, rather than complementing social time. Location and timing of internet use determines which interpersonal relationships are affected.

4.3.2 Recreational Reading: The studies of Mokhtari, Reichard and Gardner (2009) reveal that students spend more time on an average on internet as compared to that of academic reading, television watching and recreational reading as more number of students enjoyed using internet than doing other activities and they were more likely to use internet almost every day.

4.3.3 Time Displacement: According to Mokhtari, Reichard and Gardner (2009), internet has emerged as an exceedingly prominent new information and communication technology that might interfere with the time that students spend in doing recreational activities and thus marks the concept of displacement hypothesis that is based on the premise that internet has created a shift in people's allocation of time during the day, i.e. more time on the internet equals less time on other activities. It explains that internet use is associated with reduced amounts of time spent on socializing with family, friends, and colleagues.

4.3.4 Time Creation: Mokhtari, Reichard and Gardner (2009) propose another hypothesis called *efficiency hypothesis*, which suggests that internet serves as an added resource, makes users more efficient and results in less time spent on other activities. The findings showed that internet appears to be a powerful informational resource for respondents who spent more time on internet as compared to other activities and marked that people enjoyed using internet for e-mail, chatting, surfing, conducting research and listening to music. The study concluded that the time college students spent on the internet did not appear to displace or interfere with the time they reported spending on reading for academic or recreational purposes, thereby lending support to efficiency hypothesis.

4.3.5 Newspaper Reading: Vyas, Singh and Bhabhra (2007) highlight that the three key elements that drive the newspaper industry are content, delivery and advertising, which internet is very well capable of providing at much lesser cost. In addition to this, it also provides convenience of reading news as per discretion and availability of time and cost advantages for those who have free internet access and allows for simultaneous scanning of two to three similar news sites to find out different opinions on the same issue.

Also, the amount of information available is abundant and easily accessible from anywhere. The analysis indicated that more updated and multiple reporting, for the newsreader on internet highly correlates with perfection dimension; presentation, amount of news and availability with access dimension; and convenience and cost advantage with value dimension. The authors found that internet has resulted in time displacement of newspapers and is more popular in case of business/stock news, sports news and international news. It was concluded that reading news on internet is complimentary to reading newspaper and it is not substituting the newspaper.

4.4 Internet for Research and Academics

4.4.1 Worldwide Connectivity: Perry (1995) explains that internet is a global network of networks that connects more number of people each day in different countries worldwide to more than a billion computers. It adds that the nature of the internet creates an environment in which information is shared with colleagues in cross-disciplinary research that spans across the globe.

Parent and Cruickshank (2009) also validate that internet fosters the creation of virtual communities of individuals who are geographically distant but linked with common interests and concerns.

4.2.2 Information Sharing: Chen et al. (2009) report that internet provides information originating from governments, institutions, corporations, or individuals for a wide array of purposes that include research, learning, business, politics, charity and leisure in various forms like e-mail, web pages, blogs, wiki resources, instant messages, open access journals and books. It is for these reasons, that users consider it as the first choice for information search.

According to Parent and Cruickshank (20090, internet has highly influenced the way information is accessed and shared around the world.

Zumalt and Pasicznyuk (1998Z) apprise that internet has an ability to store billions of records such as complex data sets, high-resolution graphics, sounds, videos and animation that are available in addition to plentiful of databases and documents.

4.4.3 Academic Use: Applebee, Clayton and Pascoe (1997) note that internet has made a significant impact on academic work as it has resulted in improvement of productivity, quality and efficiency.

Furthermore, Dowling (1998) adds that the complete accessibility of educational resources via the internet generates consequences in relation to knowledge creation and raises concerns as to whether the timeframes traditionally associated with research degrees should be reduced in acknowledgement of benefits that arise out of more proficient information retrieval and efficient effects of word processing on academic writing. This development of technology has resulted in reduction of social complexity that is innate in traditional system of interaction and has maximized benefits for academics.

Moore (2000) maintains that e-books and electronic journals are available on internet which can be accessed freely whereas others ask for subscription charges.

Parent and Cruickshank (2009) verify that the invention of World Wide Web provides a coherent view of a library of files that are distributed across host computers all over the world. In addition to this, hyper-linking has also allowed them to follow the thread of one file on one computer to a related file on another and thus creates an increased demand for means of information search.

4.4.4 Reference Tool: Devlin and Burke (1997) describe internet as a very useful reference tool as it includes documents in different media, both hardcopy and electronic, besides non-conventional sources of information such as information centres, individual experts and electronic mailing lists. It adds that categorization increases the usefulness of internet as a reference tool as it would help in matching particular resources with the specific question types.

Zumalt and Pasicznyuk (1998) confirm that internet has the ability to answer wide range of questions and is a stable reference medium as internet based results can be verified with a traditional reference source. The authors note that many research professionals use it as an instrument of choice rather than a tool of last resort.

However, Chen et al. (2009) note that web information is not regularly cited as reported in a number of publications due to lack of confidence among researchers in the quality and reliability of resources.

4.4.5 Supports Collaboration: Perry (1995) proposes that the interactive nature of internet plays a formative role in the discussion of original results or findings and provides a venue in which scholars from around the world can participate in the reasoning and sculpt the information into a larger and cohesive whole.

Dowling (1998) adds that internet has the ability to support collaborative research and writing as it facilitates exchange and modification of textual material by writers who can interact through a range of software types and makes assembling of electronic textual material a reality.

Moore (2000) advises that development in electronic resource purchasing has resulted in a decline in possession of print items, which further causes libraries to explore co-operative collection and development of print materials to ensure access and maintenance.

4.4.6 Plagiarism and Copyright Issues: Dowling (1998) highlights that the ready availability of massive amounts of material in useable form significantly facilitates the problem of plagiarism. It gives rise to copyright issues as it seeks the prior consent of original author/owner of electronically mediated textual information.

Parent and Cruickshank (2009) also confirm that internet age gives rise to issues of copyright, freedom of access to information and enfranchisement.

Chen et al. (2009) add that research information on the web comes with varying quality, as some information is available without any name or affiliation, which is a major determinant of reliability of information.

4.4.7 Virtual Library: Moore (2000) describes virtual libraries as libraries without walls and informs that internet provides the windows and transparency to these libraries. It allows users that are physically isolated from the library to see in and those who are inside the library to see out. It reassures that internet plays an inevitable role in virtual libraries as they have become areas of core importance for distance learning programs. The widespread access of internet into homes of students ensures the success of these schemes and facilitates the libraries to offer course materials and other documents electronically to students scattered over wide areas, particularly rural areas.

Parent and Cruickshank (2009) affix that just as libraries

are getting transformed into databases of digital objects from being brick-and-mortar structures, librarians are also getting transitioned from being custodian of books to being leaders and enablers of the information. Libraries now are like internet cafés, satellite-equipped bookmobiles, a collective with a laptop, a person with a cell phone and thus can close the informational gap that exists between skilled and the new internet user. The librarian needs to be a teacher of and guide to the resources available, and also an advocate to ensure that these resources continue to be available and new ones are continually brought forward.

5

Internet and Social Capital Formation

5.1 Social Interactions

Alrayyes (2012) recognizes the scarcity of time in present and future lifestyles that have made internet being accepted as an inevitable medium that connects people to each other, to make lives more easy and productive.

Lee, Leung, Lo, Xiong, and Wu (2011) apprise that internet acts as a mediator in broadening people's social experiences and involvement with family, friends and peers by providing communication solutions, helps in bridging the physical distance among people and brings them closer emotionally as well as culturally. Ease of connecting with people through internet has transformed it into a *hospitable* place and many people use it along with traditional ways of communication. Internet has not displaced people's involvement in communicating with each other, but it has very much displaced the way it used to be done.

Brignal III and Valey (2005) notify that individuals online are more likely to respond in a quick and spontaneous manner, as there is no shared physical space to disrupt, no implicit social contact, and few social interaction rituals to prevent individuals from being rude when delivering their responses.

5.1.1 Relation between Virtual and Physical Spaces: Alrayyes (2012) argues that virtual and physical spaces co-exist due to their strong interconnection and people like to experiment with what they do in virtual spaces in reality. The author supports this argument with following three examples. First, companies advertise and engage their target customers through virtual spaces and drive them towards their physical stores like what has been done by Angry Birds. Second, in music industry, virtual spaces have allowed the conversion of physical products into softcopies and resulted in market expansion, sales growth and reduced production cost, but, undoubtedly, physical spaces are used as venues for music concerts that are marketed virtually as virtual spaces are used to spread the word, expand audience internationally and increase popularity. Third, the development of social networks reflects demands of social democracy as they have become the spaces where people gather in to discuss their demands and organize protests like the use of Twitter during Egyptian revolution, when it was used to connect people, share latest updates and venue for next gathering and acted like a walkie-talkie to share information of what the government was doing to stop protests.

5.1.2 Changing Structure of Social Gatherings: Alrayyes (2012) conveys that virtual spaces help people to share their interests and social needs and thus lead to creation of online communities that allow public to join activities virtually and really. He adds that the change in culture of recording music affects city and physical settings as in restaurants, pubs, and cafeterias, the physical seating arrangements are decided on how people listen to music—either they share or they may listen to it in isolation through their personal headphones/earplugs.

Brignal III and Valey (2005) reinforce by adding that online social support offers benefits that face-to-face social networks cannot: anonymity, constant access to better quality expertise, and enhanced modes of expression, with less chance of embarrassment and without incurring an obligation to the support provider.

5.1.3 Displacement of Social Ties: According to Lee, Leung, Lo, Xiong and Wu (2011), internet has led to time displacement—using the internet has displaced time spent earlier on engaging in social and outdoor activities and acknowledge that interpersonal communication through internet helps in forming core and significant ties that help in networking. They add that internet has enabled people to displace poor relationships with better ones through increased social involvement and well-being; results in increased strength in relationships with family and friends due to increased events of working at home.

5.1.4 Level of Social Connectedness: Taube (2004) opines that the dispersal of new information and communication technologies will affect social life strongly and the amount of actual social relations of a person will diminish with increasing use of such technologies. The study identified that internet causes increasing disconnectedness of society as time that was previously spent in social interactions has been reallocated towards internet use. He further adds that internet has also led to decrease in social contacts, as it can be the ultimate isolating technology that reduces an individual's contribution towards the society.

5.1.5 Impact of e-WOM: Cheung, Lee and Rabjohn (2008) explain that electronic word-of-mouth allows consumers to obtain information related to goods and services from a vast, geographically-dispersed group of people who have had experience with relevant products or services. The authors imply that information transfer through e-WOM may vary from person-to-person and same content can incite very different responses in different recipients. The study concluded that relevance and comprehensiveness of online opinions are key elements in decision-making process and have significant impact on information adoption and usefulness.

5.1.6 Difference between Face-to-face Interactions and Internet Communication: Brignal III and Valey (2005) inform that individuals in face-to-face interactions give out more subtle cues such as posture, hand gestures, tone of voice, movement in a conversation, eye contact, and levels of social formality that are generally absent in online interactions and argue that limited capacity to detect such cues inherent in computer-mediated communications has negative consequences as it gets impossible for people to change the course of their conversation to relieve stress or reduce emotional discomfort. Furthermore, it could undermine the perception of leadership, status, and power, and leads to reduced impact of social norms and therefore to deregulated, anti-normative behaviour.

Lee, Leung, Lo, Xiong and Wu (2011) showed that interpersonal communication via internet has negative impact on people's quality of life and cannot replace face-to-face conversations in improving relationships and suggests that people who have offline social support tend to enjoy face-to-face communications more than online interactions. Socially isolated people, due to absence of non-verbal and emotional information that impairs their communicating ability, are less likely to indulge in internet communications.

5.1.7 Impact on Young Generation: Brignal III and Valey (2005) notify the point that the heaviest users of internet, i.e. the cyber-kids who spend most of their time on computers and internet for various activities and spend less time having face-to-face interactions with peers may have significant consequences for their development of social skills and presentation of self due to absence of visual and auditory cues that are present during personal forms of interaction. The authors highlight that social freedom leads to emergence of a *cyber-self*, wherein teens manipulate their online identities, pretend to be different people, give out false information to people when they communicate online and this helps little in the practice of maintaining stable relationships. They add that children and teenagers who are heavily involved in communicating through internet could be improving their imagination and cognitive abilities but are also evading reality by avoiding necessary practice of social interaction rituals. Instead of learning the interaction skills, they are learning how to deceive individuals and they may not be able to trust accurately without visual verification through personal modes of interaction.

They warn that if children develop interaction skills based on online communications and without enough experience of personal interactions, the likelihood of friction increases. It can be argued that new cultural and social phenomena have typically produced tensions between the generations as internet environments are training students to work as individuals rather than as members of any social group, where people can choose with who they wish to interact and how they want to manage it.

5.1.8 Negative Aspects: Brignal III and Valey (2005) interestingly note that online behaviour is different than that of offline behaviour, as people are willing to misrepresent themselves by feigning a different gender, skin colour, sexual orientation, physical condition, or age; open display of group norm violations such as aggressive behaviour, racism, sexism, homophobia, personal attacks, harassment; and a tendency for individuals to quickly abandon groups and conversations, refusing to deal with issues they find difficult to immediately resolve. They add that impersonal interactions through internet environments have made people employ multiple roles in their interactions with others, which are chosen for employment on the basis of audience segregation.

5.1.9 Better or Worse: Brignal III and Valey (2005) argue that internet will revolutionize social interactions, politics, communities, or society because it allows ideas to circulate to a wide audience and thus helps entrepreneurs with good ideas to find capital and bring expertise to market products and services. It also helps flatten hierarchies, dilute power and reduce monopolization of information, permit new and interesting forms of community, make citizen activism easier and encourage a self-reflective society.

5.2 Internet and Teenagers

Teenagers have taken to internet like fish to water. Like water is lifeline of a fish, similarly internet has become no less than a lifeline for modern day teenagers. Gencer and Koc (2012) have recognized that the adoption of digital technologies is higher in young adolescents than adults. New products, communications and advertisements are designed in a way to appeal to the youth culture, as they are the centre of lucrative digital marketing enterprise. Due to the availability of high-speed broadband, young people have been the main beneficiaries of easy and fast access to internet and spend hours being online each day.

5.2.1 Create Own World: According to Choi and Ross (2006), internet has granted the power to create *own world* to the youth, where they are allowed to talk freely. They use internet in a variety of ways that spans from educational purposes to entertainment. Gencer and Koc (2012) support this argument by mentioning that the increasing scope of uses of internet interacts with people's needs and motivation and leads to multiplication of inevitable and well-evidenced advantages of internet. Choi and Ross (2006) also acknowledge the fact that websites are becoming online communities; making it possible for young people to do everything at one place and in order to keep customers glued to them, they lure them into upgrades and make them continue to pay for virtual goodies through real money that is funded by parent accounts. Lee (2008) noted that digital expertise among young people has revealed in a way that it has led to application of new technologies, with many having their own online content in the form of blogs and personal sites.

5.2.2 Safety Valve: Internet acts as a safety valve for youngsters from the demands of their parents and pressures of succeeding academically. Internet has made way for them to gain popularity and hold dominant status amongst a group of friends and satisfy the desire that they were not able to do so in their academic endeavours. Online gaming, information and data sharing could allow them to escape from reality and be a part of the competition. Winners and leaders were blessed with influential and commanding powers, whereas people who lacked these skills found themselves isolated and victimized. The fear of being on the losing side makes youngsters invest more time to maintain the status of their character.

5.2.3 Erosion of Parental Control: Choi and Ross (2006)

narrate that the use of internet by children was originally initiated by parents as they believed that it will ensure better academic performance, but they failed to realize when internet was being used in different ways for many other purposes like access to pornographic sites and the promotion of violent behaviour. Lack of knowledge about internet and its full potential among parents made it possible for youngsters to conceal the extent and type of their internet activities and such disobedience has led to violation of parental authority.

5.2.4 Negative Aspects: Choi and Ross (2006) stress that the disturbed and increased use of internet has highlighted concerns in terms of anxiety about health, interpersonal problems, declining academic performances and dilution of traditional family values. Gencer and Koc (2012) support this view by adding that internet, as a strong source of need satisfaction, can even defy the intimate connectedness of collectivist cultures and generate isolation in strong family ties. Excessive use, intolerance, preoccupation with the internet and use of internet to escape negativity are some of the symptoms of serious disturbances in user's social capital.

Lee (2008) apprises that internet access from home allows youngsters to use more internet and access a broader range of content, which cannot be done in schools due to restrictions that govern the use of internet.

5.3 Search for Partner

5.3.1 Efficient Search Tool: Rosenfeld and Thomas (2012) advise that internet search is important in finding potential partners for individuals. Bellou (2014) briefs that searching for a suitable partner in an offline and decentralized environment can be a lengthy process, while an online centralized market could resolve a number of issues as it allows for targeted search with certain desirable characteristics with users holding some degree of secrecy. Internet could be professed as a way of declining search frictions by reducing the cost of search and increasing the offer arrival rate.

5.3.2 Meeting Point: Bellou (2014) apprises that now-a-days more people access internet than before and can use online resources as a fast and possibly more proficient way of meeting others via instant messaging, email, online dating and chat rooms. Internet is one social arena that is unambiguously gaining importance over time as a place where heterosexual couples meet.

5.3.3 Social Intermediary: Rosenfeld and Thomas (2012) advise that internet plays the role of a social intermediary as it fosters new social relationships between two strangers. Internet compresses social world and allows people to find each other and meet without any other intervention. Bellou (2014) further adds that internet plays an important role in family formation as it is being used by people as an intermediary for partner search.

5.4 Political Participation

5.4.1 Abundant Information at Low Costs: Getting information has cost implications. Internet has the potential to reduce these costs that will lead to increased participation and thus help people in becoming better citizens. Internet's ability to easily multiply and amplify an individual's information gathering abilities coupled with its interactive nature makes it a promising tool for overcoming time-costs associated with political participation.

5.4.2 Impact of Social Media: Fu and Chou (2016) indicate that social media can bring the dropping performance of long-ruling political party into limelight, lowering public support for them, which allows adherents to play more active role and change the issues under discussion in an election. Social media has also made it possible for the internet to become a new medium for political campaign and fundraising. Websites and blogs allow people to express opinions, diversify their political views, discuss and interact with politicians directly. Internet allows users to target their exposure to preferable information and like-minded individuals, promoting network building and knowledge homogeneity enabling such groups to stand for what they believe in and enter into a conversation/dialogue with those having a counter view.

5.4.3 Increased Real Participation: Ikeda, Richey and Teresi (2013) express the opinion that internet may be keeping people at home by giving them tools for virtual communication which may reduce highly valued voluntary experiences. Jennings and Zeitner (2003) support this view and add that growing and widespread use of internet is alleged to threaten face-to-face communication that characterizes volunteering. However, internet leads to replacement of spectator activities with more active forms of behaviour.

5.4.4 Impact on Other Media: Jennings and Zeitner (2003) highlight the fact that internet has been posing noticeable threat to pre-existing media as people prefer to use it to follow public affairs, politics, and news as compared to newspapers, magazines, radio and television or use it as an additional supplier of information. We can say that internet has reduced the significance of other media.

5.4.5 Overload Leads to Mistrust: Jennings and Zeitner (2003) raise the concern that physical isolation associated with internet use as well as widespread instances of uncivil net behaviour can drive down levels of social trust, which are crucial for people to work collectively for a common cause. They further speculate that internet via its chat rooms provides an ideal forum for fostering and abetting political cynicism.

5.4.6 Digital Divide: Jennings and Zeitner (2003) argue that the *digital divide* that is associated with differential access to the internet could get exaggerated and may reinforce existing inequalities in civic engagement as people who already have resources and skills, will simply adopt internet as another tool. Heightened internet use is either related with reinforcement or widening of civic engagement gap between those with and without access.

5.4.7 Mobile Divide: Ikeda, Richey and Teresi (2013) confirm that browsing platforms have differential impacts on participation. Mobile divide occurs because mobile internet usage does not facilitate social networking and political participation the way PC internet usage does. Mobile users have started to think that there is an app for everything and search for it even before browsing it. Mobile browsing has been designed to avoid cumbersome browsing, limit internet usage, reduce exposure to news, which can serve as a source of unbiased political information and can potentially eliminate the need of full-scale browsing. Mobile use of internet being more focused on one-to-one communication makes access to search engines, chat rooms and blogs nearly impossible unlike that of a PC which can connect diverse virtual social networks.

5.4.8 Increased Participation: Ikeda, Richey and Teresi (2013) claim that by promoting virtual communication, internet can actually increase participation in voluntary organizations and politics. Further, Jennings and Zeitner (2003) propose that internet will enrich community life, strengthen civil society and democratic politics as it expands the opportunities for communication and mobilization. However, internet acts as a depoliticizing medium by deflecting people from civic matters as it engages them with its extensive use of e-mail, shopping and entertainment platforms.

5.5 Internet for People with Special Needs

Reeves (2000), proposes that HIV-positive individuals have to cope up with not just the physical consequences of illness, but many social, emotional and psychological issues also. They seek emotional support in dealing with psychological stress that comes along with this disease and can be accomplished through a network of friends, family and community-based service organizations but, due to their circumstances may not be able to get sufficient social support. Affected individuals generally lose their social skills because they stop going out, but through internet, they receive validation for their feelings and thoughts.

Bannon, McGlynn, McKenzie and Quayle (2015)

highlights that it is sensitizing to know that young and adult people with special needs, autistic disorders, intellectual disabilities, mental health problems, social, emotional and behavioural problems and specific or more general learning difficulties experience a range of difficulties in development of identity, competence and social connectedness. Internet may serve potential benefits to conquer these difficulties and the use of internet will allow them to engage in activities which support the development of identity, competence and a sense of connectedness and belonging within a social network that is essential to healthy development and psychological well-being and areas that present a challenge for this population in offline environments. The results of the study underscore that participants felt that unlike other people, internet is always there no matter what time or day it is.

5.5.1 Online Identity Development: Bannon, McGlynn, McKenzie and Quayle (2015) present the benefits of using internet in identity formation for such persons with two varying options. On one side, it is believed that identity experiments online will result in individuals feeling less sure of themselves and therefore hindering the process, potentially resulting in poorer outcomes. On the other side, it is believed that experimentation of identity online, where individuals may pretend to be someone else, can support and foster identity development as role-playing allows individuals to control how they are seen by others. It is justified to say that internet, as a mode of online interaction, facilitates the development of identity as it allows for more impression management than in face-to-face interactions.

5.5.2 Availability of Information: Bannon, McGlynn, McKenzie and Quayle (2015) put forward the idea that internet makes itself a place for self-help materials and web-based information that can help people live a normal life. Reeves (2000), too has reassured that internet promotes empowerment through providing all the relevant information one may need. Grimaldi and Goette (1999) also suggest that internet plays an important role in the lives of people with disabilities as it allows them to access e-mail and wealth of other information without having to leave home.

5.5.3 Social Support Network: Reeves (2000) maintains that internet with its wide array of functions provides a viable alternative for individuals who cannot or who elect not to pursue traditional avenues of social support and can be used for information, communication and interaction. It is a medium of connecting people who identify with one situation and makes it easy to talk about heavy and personal stuff. Internet augments social support and becomes integral to the survival of such people by providing them with an ability to communicate and have support network online.

5.5.4 Empowerment: Reeves (2000) in his study to understand the impact of internet on HIV-positive individuals realized that it gave people a sense of power that allowed them to cope with a strengthened position rather than being a victim or in a weakened position and boosted confidence among affected individuals making them feel that they were capable of accomplishments. It was noted that it empowered people to identify the virus and build up their immune system by highlighting it as an enemy with the tactics it uses to weaken affected human bodies and made people realize that they are in-control of their bodies and can be productive. Asynchronous communication on the internet allows for online learning forum where young people can take more time to process information presented before forming a response.

5.5.5 Building Competence and Self-confidence: Bannon, McGlynn, McKenzie and Quayle (2015) argue that internet has considerable potential forum for allowing young people to learn and develop competence because it allows for continued learning online and may benefit individuals who may otherwise find a face-to-face learning environment difficult; such as people who experience anxiety or have communication difficulties. Internet may also be helpful for individuals with heightened anxiety in comparison to the general population, as they may find it easier and less threatening to participate in online learning. It is also highlighted that internet use allowed the development of a range of skills including cognitive, social and communication abilities and it may also offer greater opportunity for this group to participate within a wider social community, thus, benefiting social competence.

5.5.6 Psychological Well-being: Bannon, McGlynn, McKenzie and Quayle (2015) opine that development of skills and abilities through internet use helps in raising self-esteem and makes people feel good about themselves. The authors note that internet use has a positive impact on psychological well-being in people with special needs as it offers them opportunities to make social ties and is beneficial for people who are considered to be less competent.

5.5.7 Facilitating Help to Others: Reeves (2000) found out that internet helps people to be of service to others, through opportunities such as volunteering and speaking to groups and serving as counsellors in wellness clinics for other people. It could allow people to create websites and blogs to disperse helpful information and resources to people where there is scarcity, as they realize the importance of sharing. The respondents realized that helping others also added to positive energies, eventually, helping themselves and was like a self-fulfilment.

5.5.8 Clinical Assessment and Treatment: Reeves (2000) advises that it is due to internet that people with special needs do not just take passive recommendations from their doctors, but play an active role in standing up for themselves, as it has made them being an equal and active partner along with their physicians in healthcare decision-making. It implied that internet is a valuable resource for HIV-positive individuals and thus there is growing interest among healthcare professionals in creation of programmes that alleviate stress and nurture the coping efforts of their clients who are struggling to live with the HIV. The role of internet is of high importance for clinical assessment, as appropriately adapted online forums can provide support and counselling and people find it easier to interact socially and openly online.

5.5.9 Unrealized Potential: Bannon, McGlynn, McKenzie and Quayle (2015) admit that there are possible difficulties that people with special needs may face that inhibit them from taking full advantage of the internet. These include experiencing confusion, cognitive deficits relating to processing speed and selective attention, attention and concentration problems and heightened distractibility. It has been explained that in the internet environment, generally there are lots of sources of information, including auditory and visual stimuli. Perhaps, all being presented at the same time, from various sources make it difficult for such individuals to use internet to support their special needs.

5.5.10 Negative Aspects: Internet allows users to put up inaccurate information on their social profiles. Abundance of information on internet can be frustrating, because it is difficult to identify which one is accurate and reliable. The biggest problem is locating valuable information that is available on internet.

6

Impact of Internet on Industry, Business and Employment

Internet services are regarded as catalyst in the growth of an economy. Since the information system of a country is interconnected, the development and deployment of information and communication technology (ICT) is crucial for expanding the knowledge base and increasing the productivity. Development of technologies and improvement in telecommunication infrastructure has enabled loosely bound business entities to collaborate and provide a number of services, thus connecting more communities socially as well as technologically.

6.1 Industry and Services

A study by Grimes et al. (2012) suggests that productivity increases by 7-10 percent when broadband connections are used. Another research study explored the impact of ICT on productivity of small firms. Broadband internet connection is a powerful tool for small and medium enterprises (SMEs) as it provides them connectivity to the global markets at a much lower cost than ever. It also acts as an enabling technology that allows them to adopt a number of valuable complimentary applications that may increase their efficiency. The use of ICT in service sector SMEs enables them to reach distant employees and co-ordinate production activities more efficiently. SMEs in the manufacturing sector benefit from the supply chain and customer management applications, whereas advanced communications applications are more useful for service sector firms. Some other applications like human resources, administration applications and management systems applications etc. are useful to most firms as these are not industry specific (Massimo G. Colombo, 2013).

Gains from broadband on businesses can be classified into direct and indirect. Direct gains are due to increased efficiency which results in increased productivity. On the other hand, indirect gains result through positive externalities. Broadband has a positive impact on GDP growth and increased exports, which leads to increase in the intermediate demand and growth of support sector.

Broadband technology improves productivity by facilitating the adoption of more efficient business processes. Also, extensive deployment of broadband accelerates innovation. Another way through which broadband brings in efficiency in the business is through a more efficient functional deployment of enterprises by maximizing their reach to labour pools, access to raw materials and consumers and streamlining of supply chains. In addition to the impact on productivity, broadband technology also impacts business expansion, product innovation, and new business creation. It was found that in the manufacturing sector, firms with internet access enabled by broadband, generate 6 percent more foreign sales than the rest, while in the service sector, the impact on exports was found to be between 7.5 percent and 10 percent. Besides increasing exports, it was also found that broadband has a positive impact on the development of new businesses. This results from the network effects of connectivity. When a large enough number of households are connected to broadband, the incentive to develop new businesses increases, especially around information search, advertising and e-commerce.

The impact of internet on some of the industries is briefly discussed below.

6.1.1 Power: IT-enabled power distribution systems help distribution companies in reducing aggregate technical and commercial (AT&C) losses and in monitoring the feeders online. These systems benefit consumers as well. They can pay bills online, apply for new connection and know the status of power supply. Under the Integrated Power Development Scheme, the Central Government aims to save as much as

₹ 10,000 crore annually by using IT-enabled technology in power distribution in urban towns by January 2019.

6.1.2 Internet of Things (IoT): IoT is one of the latest developments in the field of ICT. IoT basically means "anything that can be connected to the internet will connect to the internet" (James, 2016). It is a concept of connecting all our personal digital devices through a single network. It is the future of technology and a step towards smart homes and offices.

6.1.3 Insurance: The nature of competition in the sector has been evolving everyday because of changing behaviours and technological innovations. Digitalization has the power to reshape the industry and disrupt established value chains and business models. The disruptions caused by innovation of digital technology would affect the insurance sector tremendously. Driverless vehicles would alter the need of auto insurance. Connected homes would transform house insurance, cyber security and drones would create demand for new forms of coverage, and shared businesses would change the underlying need for insurance.

Claim processes are getting transformed because of automation, analytics and consumer preferences. Connected technologies allow for self-diagnose and auto-reporting of incidents/problems by insured devices. Self-service claim reporting technologies can create fast and seamless customer experiences. This has also made damage assessment quick, safe and cheap. Digitalization of insurance sector would substantially reduce human service costs, enhance customer experience and increase profitability and growth rate.

6.2 Consumer Centricity of Business

Bhatt (2015) reviews the present scenario and firms' attitude towards the benefit of all the stakeholders, not just shareholders of the company. Stakeholder engagement has evolved with the changing role of corporations in the society. It has changed from being a reaction to customer problems to their greater participation in the value chain. The stakeholder engagement has benefited many companies through informed decision-making by generating business intelligence. It also brings diverse perspectives together for creativity and innovation.

Earlier, through traditional means of marketing, there used to be a monologue between customers and businesses and a wide range of advertisement was broadcasted. But the modern customer has very less time to watch all these advertisements and therefore chooses the message he/she wishes to see. As internet facilitates interactions, customers want that companies should listen to them and respond. Technology is serving in a big way by helping companies get insights about their customers. Companies are offering customized products and services to their customers as they are no longer the passive players in the market. They already have a lot of information at their disposal. Therefore, being customer centric is the only way to survive in the modern day markets.

Kanojia (2015) discusses the concept of d*igital customer engagement* which emphasizes the dependence and reliability on customer generated content. Online modes of consumer engagement are gaining popularity as they involve internet marketing, social media marketing, blogs, web portals, crowd sourcing, e-mail campaigns etc. and have wide number of applications in different sectors. Many private companies are now using social media platforms, mobile devices and other technologies to get closer to their customers because of their volume and speed to reach customers.

Technology has also changed the way customers contact the companies. They seek automation, self-service and personalization. The technology also brings good prospects for the government and the jurisdiction of the country as it helps them to empower the citizens and gives them more access to the details through the use of location-based services. It helps in achieving long-term operational efficiencies. In the education sector also, MOOCs (massive open online courses) have been successfully implemented by the top universities of the world because of the availability of internet services.

As mobile phones (now mostly smart phones), have become an integral part of our day-to-day life, there is a need to change the way businesses reach and interact with its customers. Kanojia (2015) asserts that smart phones act as primary internet devices and thus can be used for mass marketing to gain more participation. He is of the view that to ensure deeper reach, the focus must be on developing customer-centric applications. There is a need to make these applications available in different languages to avoid the gap that may generate. Some touch points that can be of core importance for the businesses to engage with customers through mobiles are product discovery platforms, search engines, customer care portals, consumer grievance forums and digital payment mechanisms.

The proliferation of social media and the video invasion are instrumental for building strong relationships with customers. Videos ensure attention for longer duration and provide opportunities for emotional congruence. Social media relationships built on videos and social interaction, results in self-efficacy and positive word of mouth. Social interactions and shared experiences increase the levels of commitment and engagement with the businesses. Increased interaction results in more satisfied customers and thereby positive actions. To attain higher levels of consumer engagement, businesses must integrate the digital channels with traditional brick and mortar.

Finally, to be a customer-centric company, a collective effort by different departments of the organization is an imperative practice. It is unlike product-centric approach where the functions are compartmentalized. Linkages have to be developed across the organization and a better understanding of customer needs has to be discovered at every level. These linkages are being made with the help of internet technologies. Kanojia (2015) argues that even customers should be treated as business partners and their feedback and suggestions should be taken into consideration. E-tailers must first ask consumers what they want to buy rather than trying to push what they want to sell.

6.3 Marketing and Advertising

Companies use various marketing tools to increase customer engagement and participation. McGaughey and Mason (1998) examined the moderating impact of internet on consumer behaviour and role of internet as a marketing tool. Internet has become much more user-friendly with the passage of time. Online business activities are a rapidly growing phenomenon. Internet has proved to be a powerful medium for obtaining and distributing information to help marketers take product decisions and devise appropriate marketing plans.

As the popularity of internet has increased, the popularity of sports has increased as well. Viewers are no more glued to their TV screens, wishing to watch only live shows. They want the convenience of watching their favourite matches whenever and wherever they want to. Digitization of media content, proliferation of IT-enabled devices and easy availability of high speed broadband connections has given birth to the phenomenon called *media convergence*—a phenomenon involving the interconnection of information and communications technologies, computer networks, and media content. Various forms of media such as TV, social platforms and even out-of-home (OoH) advertising, are being used by brands to bridge the gap between online and offline resources. It helps them to efficiently identify their target group of sports fans.

Sports fans are some of the most loyal customers because they are open to use the products of the brands, which are participating in their favourite games. Brands not only need to provide relevant content and enhanced experience to the sports fans, but also engage them sustainably with the brand. Programmatic technology is the need of the hour. Here, with the help of triggers, advertisements and content can be displayed to the viewers based on monitoring real time developments as the game progresses.

OoH advertising is an old form of advertising which was assumed to be restricted to hoardings and billboards on roadsides. They were more of a distraction rather than attention gainers. Now, IT has shown the way to modernize the OoH advertising in the form of digitized and interactive billboards. OoH advertising has shown good results in rural markets as well. In rural areas, people gather at public places such as bus stops or marketplaces, where many OoH exercises like permanent billboards are done. Even though the quality of those billboards is very basic, they have proven to be better communication tools than most others.

6.4 Online Shopping

A study by ASSOCHAM and Grant Thornton (2015) revealed that online shopping via mobile phones is gradually increasing in India.

6.4.1 Risk and Trust: Trust plays an important role in e-commerce activities. Grabner-Kraeuter (2002) reported how trust is an important factor for the circulation and acceptance of e-commerce. There are two types of uncertainty, namely system-dependent and transaction specific. Online retailers are finding it increasingly difficult to sell their goods. Internet users are unwilling to shop online because of fundamental lack of faith that must exist between business and consumers. Trust is not only a short-term issue but the most important long-term barrier to realize the full potential of e-commerce. Consumers foresee numerous risks while buying on the internet as it does not involve simultaneous exchange of goods and money, because of the temporal separation between exchange partners.

System dependent uncertainties arise because of potential technological problems and lack of clear legal norms. On the other hand, transaction-specific uncertainties arise out of asymmetric distribution of information between the transaction partners. Trust plays a key role in buying process as consumers tend to look for experience in credence qualities of products and their sellers. Trust functions to lower the perceived risk of a transaction. On the other hand, there are some risks that come about during exchange of information via internet.

System-dependent uncertainty comprises of events that are beyond the direct influence of actors and can be characterized as environmental uncertainty. In the context of the internet, it relates to potential technological sources of errors and security gaps.

Technical safety gaps can emerge either in the data channel or on the final points of the process like the desktop system of the customer, the server of the internet retailer or the involved banks. Transaction specific uncertainty may arise out of decisions of economic actors and is caused by asymmetric distribution of information between the transaction partners. This primarily concerns the quality of products and services offered on the web because the quality assessment in electronic markets is much more difficult due to disappearance of the personal factors that are used in traditional modes of selling. This type of uncertainty can be reduced by application of concepts of search, experience and credence qualities. Online product evaluation is possible through performance-oriented information substitutes that relate to the seller and his reputation in the market or adequate indicators such as brand names or performance bonds.

Online retailers facilitate the evaluation of credence qualities by signalling trustworthiness like referring to a certificate of a trusted third party. Trust is a potentially important aspect for reducing uncertainty and complexity of online transactions and relationships. Trust is considerably useful in reaping economic benefits. Because in the near future trust will remain the decisive factor for success or failure of e-businesses, it is very important for internet companies to act in a way that engenders consumers' trust. Efforts to increase the security of e-commerce systems and trustworthy behaviour of online retailers will prove to be of advantage for both consumers and companies engaging in e-commerce.

6.4.2 Nature of Online Shopping: Brian T. Ratchford (2016) examined the determinants of the use of internet as a source of information on automobiles and its impact on the use of other sources, and on total search effort. The study found that the share of search time spent with the internet is inversely related to age and positively related to education. Younger and more educated consumers are significantly benefited by the internet. People consider internet as a source of information. Consumers use internet to gather information about automobiles. Internet leads to reduced search for automobile on an average, which is more pronounced for those who are likely to use the internet, and are relatively young and are highly educated and rich. The total search for automobiles across all sources increases with gains to search, as measured by sticker price, but decreases with costs of search, as measured by log of wage, and with variables related to prior information and age, which are used as a measure of experience.

6.4.3 Websites Shopping: Ganpathy (2001) reported key dimensions of B2C websites. There has been a phenomenal growth in the number of organizations using internet for marketing, promotion, and transacting products and services because of the growth in B2C e-commerce. Firms and consumers have started using internet to gather information and for online shopping. Georgia Institute of Technology conducted a study to gauge the primary use of internet by the consumers. The study reported that 40 percent of the respondents used web for shopping primarily. The usefulness of a B2C site not only depends upon the content and design but also the navigation tools. There has been much technical advancement in internet security including cryptography, digital signatures, and certificates and authentication. Despite these advancements, consumers are apprehensive about online shopping. An effective B2C site provides quick access to information about products and services, information about the organization (the seller) and the contact information.

6.4.4 Quality of E-Tail Experience: Hans H. Bauer (2006) presented the need to integrate the serviceable and hedonic motives into one measurable scale so that the quality of e-service can be revealed explicitly. A substantial number of customers experience problems while shopping online such as lost orders, service breakdowns, or inadequate handling of complaints. All these problems make the maintenance of the quality of e-service challenging.

The level of customer satisfaction with the quality of e-tail experience and what makes a customer shop online has been studied by Wolfinbarger (2016). It also suggested various factors that predict customer judgment of quality and experience. Online shopping experience is different from offline shopping experience. There are many factors affecting online purchasing behaviour of consumers such as attitude towards technology, skills and experiences, people having greater internal control, knowledge of internet, willingness to adopt new technology or those who spend much of their time online, are more likely to shop online. Furthermore, those who are time starved are more likely to shop online. Other factors also influence the decision of shopping online or offline, such as type of product, customer mood and goal behind shopping. It is also found that website design, reliability and security/privacy serve as the basis of judgment for the quality of an online site. Research suggests that as more and more information and technological tools are available to customers, their preferences and experiences are likely to change.

6.5 Travelling Displacement

Hausmann (2016) has explained the need for business travel and identified that in the times of internet, the need for business travel may get eliminated. Internet allows business and people to call, e-mail, see and talk to counterparties without travelling to distant locations. Most of the business travel relates to management of foreign subsidiaries. The brain needs to move and not the bytes, primarily because:

- Brain has a capacity to absorb information, identify patterns and solve problems without us being aware of how it does it. It lets us understand facial expressions, body language, intonation and other subtle indicators that are gathered unconsciously and results in better evaluation.
- Brain is designed to work in parallel with other brains.

Thus, it is noted that the amount of travel is related to the amount of know-how that needs to be moved around. Thus, it is suggested that firms must try to move brains for key tasks and not just for relevant information to the brains.

Hemrick's (2004) study was done through a web survey and explores whether the use of e-mail and the internet has led to more or less trips and more or less driving during rush hour and overall. Snowball sampling method was used to determine the sample and about 118 responses were received from a 31-question survey that was administered online. The respondents were found to be active users of e-mail and internet, having ready access to them. E-mail was used to build or at least maintain social capital. The internet is mainly used as an information resource. E-commerce is also a major use, especially when making purchases, downloading software, making financial transactions, placing product service requests, and downloading music. The respondents were active drivers too, with most of them driving more than 100 miles per week.

The results suggested that e-mail and the internet use led to trip substitution as many respondents indicated that they used information technology instead of making trips to work, bookstores, other stores, and libraries. Information technology was also frequently used instead of making trips to clothing stores, music stores, friends' houses, government offices, and special interest organizations. This resulted in about 18 percent lesser trips.

Internet is used to gather information to prompt trips to bookstores, other stores, government offices, movies, libraries, and other stores. This survey supported the conclusion that use of information technology by this group of respondents led to less driving rather than more. The study suggested that information technology has a very high potential to reduce traffic both during the rush hour and overall if large segments of the population come to use e-mail and the internet.

Age was considered an important variable to describe the reduction in number of trips, probably because of immobility that comes with it. Similarly, people with less education had more time to spend on e-mail and the internet, so they could take more advantage of the technology.

6.6 Employment and Earnings

6.6.1 Help in Job Search: The impact of internet on job search in Germany was estimated by Chau (2012). The study found that internet plays an important role for the job seekers and employers as it has saved the time of both. Job seekers who use the internet have 4 percentage points higher probability of being employed in a 12-month period. University degree holders were more likely to be employed in a 12-month period. Unemployed people aged 26-35 were most likely to be employed within the 12 months period. Job search via federal employment institution had a negative effect on the search results. In contrast, job seekers who contacted employers directly had much higher probability of being employed in a 12-month period. Single women job seekers are less expected to be re-employed in a 12-month period. Job search on the internet can definitely serve as a signal. The employers and the job seekers both are proficient in using computers and browsing internet.

6.6.2 Impact on Earning: The impact of internet use on earning of US workers was examined by Bonikowski (2008). People who have access to the new digital technology have benefited in the form of improved employment possibility. Internet use is becoming necessary for certain kinds of social and political participation and for access to some private markets and government services. Technology-linked wage gains show improved productivity due to the use of the new technology at work. Internet use may also contribute to earnings by enhancing access to labour-market information and by serving as a signal of status or competence. Internet can be used as a source of social capital and as a cultural resource. Internet users may benefit from online job listings, expanding personal social network, benefit their employers with useful information and contacting clients or setting up collaborative ventures. Internet users earned more than the non-users. The labour market rewarded internet use at home and at work, and workers who went online at home and work, did best of all.

6.6.3 Impact on Professionalism: Ellen M. Harshman (2005) studied the impact of internet on professionalism. Traditional concepts of professionalism have been significantly challenged by the appropriation of internet technologies. Some people misuse internet on professional ground because its authenticity cannot be verified. The emergence of new forms of expertise on the internet, by challenging established institutional and professional guarantees of expertise, undermines important and powerful arbiters of ethical forms and behaviours. Convenience, access, and price act as motivators for seeking non-traditional sources of advice and information on the internet. It provides a sense of belonging.

An analysis of history of professional practice recommends that institutional and professional interests have colluded to create a widening net of claims for exclusive rights to the possession and distribution of information. The increasing use of internet to access knowledge and advice—once thought to be the exclusive province of particular professions—challenges both the meaning of the term profession and the infrastructure assumptions which supports codes of professional conduct. On the other hand, internet can also become a means for professions to successfully negotiate the challenges created by the virtual environment. It is concluded that the internet provides a means of reinforcing traditional professional practice, thus increasing the quest for alternative sources of knowledge and information, leading to the pressure for internet regulations on the part of the professions and increasing the quests for alternatives.

6.6.4 Impact on Labour Market: Singh (2008) examined the impact of internet and e-commerce on labour market. E-commerce has played an important role to generate employment in the economy. This is possible through improved labour facilitation and direct employment for developing countries. The maintenance and supervision of the additional hardware and software systems that are pre-requisites for e-commerce have led to the need for more computer workers. Internet technologies that drive the global information revolution that is happening around have made outsourcing of service-related jobs to developing countries possible. The rise of internet has also led to increase in women's presence in paid employment and has further reduced gender-based discrimination. Internet has positively impacted the labour productivity because e-commerce reduces the coordination costs between different work processes, facilitates firms to fragment tasks and thus enables them to improve labour productivity. Development of more interactive job search sites allows job seekers to store search preferences and user profiles and easily follow-up on identified vacancies by making online applications.

Internet is invaluable for employers as well. The ease of posting employment openings, the relatively low cost of advertising and the speed of candidates' response are all positively related to the use of internet. There is also a shift in labour demand towards high-skilled workers (i.e. IT workers) both in manufacturing and non-manufacturing industries.

6.6.5 Impact on Married Women's Labour Force Participation: Dettling (2016) examined the impact of residential high-speed internet on married women's labour force participation by using instrumental variable strategy. Educated married women with children are actively using internet and there is a large increase in their participation. Internet has saved time and money as it facilitates individual life in different ways. The high-speed internet use has positively impacted the labour supply decisions of married women as it facilitates work-family balance. There was a declining aggregate trend in labour force participation for men, single women and married women in the early 2000s. Tele-work has increased the wages by reducing travel times and by improving productivity through reductions in absenteeism. Effect of high-speed internet on labour supply will be relatively more positive for married women than single women and men if the internet is used for tele-work and home production.

High-speed internet increased labour participation rate among college-educated married women with children. Less-educated women and highly-educated childless women exhibit a similar, somewhat smaller labour-supply response to high-speed internet, while both are positive relative to men and single women. It was found that married women with high speed internet were less likely to be unemployed than single women or men who used high-speed internet.

7

Umbrella Issues in Internet

Umbrella issues encompass all the areas that have an impact on all aspects of behaviour. For instance, transportation has an impact on social capital, human capital, and economic needs as well. Similarly, gender inequality, privacy issues and laws pertaining to internet can be regarded as umbrella issues.

7.1 Impact on Everyday Life

Sung (2003) examined the impact of high-speed internet on the user behaviour in Korea. The consumer behaviour towards internet usage has changed with rapidly advancing internet network technologies. The impact of high-speed internet on consumer behaviour was measured in terms of purchasing over the internet, internet services, information exchange and chatting, online problems and user satisfaction. The internet is affecting socio-economic activities and cultural aspects of organizations and individuals. Younger generation uses internet for leisure activities, e-shopping, e-auctions and to fulfil other personal needs. E-commerce has been on a constant growth due to many factors, e.g. convenience, saving time, lower prices, easy comparison and vendor information availability.

The older group with higher purchasing power and income were the actual online buyers. Internet provides a variety of services to its consumers such as email, address book and online tax filing. Internet users can exchange any information, documents and communicate via email. Users can obtain any information with the help of search engines provided internet portal sites like Yahoo, AOL etc. The traditional mass media (television, radio, newspaper and magazines) have been replaced by internet radio, online television and e-magazine.

The study found that the current transition of network has no influence on those who are willing to participate in commercial activities on internet via e-shopping and e-auction. Network development is not a cause of any problem.

Liang (2010) highlighted the advantages of internet and how it brings about essential benefits that improve the quality of life. It was found that people are of the opinion that the use of e-services is improving their lives. Four major dimensions that were in the framework of e-life indicators were e-daily life, e-business, e-government and e-infrastructure. Internet has a significant influence on the quality of life, including social, leisure, economic, and community well-being. The internet also offers individuals a medium through which it is possible to communicate with the world around them. E-mail is also found to facilitate social contact. Online chats/discussions have a positive impact in some cohorts. Internet and email are excellent sources of support and enjoyment, resulting in improved quality of life for older homebound adults.

However, internet use leads to family alienation in adolescents. The correlation coefficients of e-service cognition and usage with quality of life revealed that usage of e-services has a strong and positive correlation with the quality of life than that of cognition of e-services. People can experience high overall quality of life when they have higher degree of cognition and deeper degree of usage. Merely hearing or knowing of availability or presence of e-services does not promote quality of life. But, their usage can bring more happiness to someone's life.

Blas (2006) analyzed the key drivers of internet dependency and its impact on willingness to purchase online. Internet has become a useful communication source. It is very beneficial for businesses and consumers. Internet has been growing fast with increasing penetration rate. Internet users are not an identical behavioural group of people. The study found internet-dependent users as the young internet users, with high educational level who feel affinity with the medium. Greater exposure to the medium is the main cause of media dependency. It has been found that some behavioural effects such as the purchase of products and services can be increased when dependency level with a medium is high. Consumer needs and interests change with age, with younger individuals being more favourable to internet.

Furthermore, lack of acquaintance with the medium prevents adult consumers from assessing the benefits of internet and they tend to prefer other information sources for their purchases such as television, press or social interaction with friends and relatives. Persons who have used internet the greatest are more dependent than those who are less well-known with the medium. Thus, greater online experience leads to better relations with the medium, as the individual is more familiar with it and values its benefits more.

Groselj (2014) examined the dimensions of internet use based on a representative sample of the population in the UK. The study used principle components analysis to identify 10 distinctive types of internet activities out of 48 activities—ranging from e-mail to online banking, to gambling and to selling products. The study described three dimensions, i.e. amount of use, variety of use and types of use along with the internet use.

Amount of use (first dimension) is a continuous variable measuring the frequency of internet use in day-to-day life. Some people use it for many hours every day and others only once a week. The amount of internet use (second dimension) is an interval variable measuring the number of different activities undertaken online. Type of internet use (third dimension) is a series of nominal variables describing different activities (sending emails to investing in stocks to making travel reservations to gambling) people engage in online.

The common internet users are doing 25 different activities that are surprisingly large in number. As internet develops, it is likely that the common activities on the internet could become one way to measure changes in the internet.

Tracey (2001) studied the impact of internet on the everyday lives of UK citizens through the integration of quantitative longitudinal time-use data and qualitative interviews. The study analyzed the causal relationships between their acquisition and use of new information and communications technologies (ICTs) and changes in their behaviour and their social, symbolic and economic capital.

The pattern of internet usage is not similar in different age groups. The youngest group spent less time using e-mail than the web. In fact, they spent less time using e-mail than the 25- to 34-year-old and 55-year-old and older groups. However, this pattern is almost exactly reversed in the oldest group (aged 55 years and older), who used e-mail more than the web. Different kinds of people make differing uses of the range of applications and services that the internet supports and probably for differing reasons. The impact of internet on everyday life in UK is not identical. The variety of activities that were being displaced comprise reading newspapers, magazines, and books, writing letters, typing on the typewriter, watching television and playing computer games. Work-related changes such as shifting from home-based to office-based work or the reverse in the case of new teleworkers or new self-employed appear to affect access to and the style of internet usage.

7.2 Gender Inequality

B. Paige Miller (2006) examined the impact of the internet on the research careers of female scientists in three developing countries—Ghana, Kenya and India. The premise of the research is that gender inequality is related to differences in human and social capital and women are less likely to acquire advanced degrees and more likely to experience educational and organizational localism. The diffusion of the internet to developing world has the potential to alleviate the difficulties many researchers experience. New ICTs enable users to access a wide variety of information quickly and relatively cheaply, as well as develop and maintain contacts with other professionals.

Face-to-face interviews were conducted with academics and scientists employed in international research centres and universities and in broad range of science and engineering fields. Five key dimensions of research careers were studied—human capital, localism, access to resources, social capital and productivity. The study suggested that the differences between educational attainment and gender were statistically insignificant. The gender gap was observed to be declining during the test period with women displaying increasing educational and organizational localism. It was noticed that the time spent in a foreign country hugely differed between the two genders. Information and communication technology has provided new means for the development of social capital as more people have gained access to it and this can be done without generating any disadvantages for women, as both of them get equal access and advantage due to internet.

Women scientists continue to exhibit localism not only in their educational and organizational environments, but also in their pattern of social relationships, which has profound impact on their productivity. The widespread diffusion of internet throughout the scientific communities of the developing world is a development that has great potential for changing the status of women. This is because access to ICT does not differ in its advantages for men and women. The increased access to ICT diminishes the necessity for travel, as it makes scientists capable of interacting with each other at a distance. There is a widespread expectation that diffusion of internet will eventually yield career gains for women, enabling them to circumvent restrictions on their mobility.

7.3 Privacy

Free public wi-fi has undoubtedly become a pre-requisite for our generation. It is the first thing that people look for anywhere they go. However, it is important for us to know that it carries certain risks that may cost us a lot. Millennials seem to be the most vulnerable to attacks as most of their work, business, and socialising is done online. Many Indians would be devastated if their personal financial information was compromised and if there is a credit card fraud done to them. Hackers and cyber criminals regularly run diet and fitness related promotional links, because many Indian citizens click onto them in order to achieve their dream bodies and end up sharing information like email address, full name or age. Various forms of attacks are man-in-the-middle attacks, sidejacking, rouge networks and malware. The biggest problem with public networks is the lack of authentication. There is not any password required to join as you would normally have at home or in private networks. This means anyone can join the network and start tapping into your information easily. There are no absolute guarantees even when precautions are in place. Thus, one must stay vigilant and restrict the types of activities that are conducted on free wi-fi to safeguard their privacy. This is one of the side-effects of living in a heavily interconnected world.

7.4 Internet and Law

Forcht (1998) analyzed the role of legal measures and regulations that determine what is acceptable and what is not acceptable, where internet has started to govern a major portion of our lives. It is of core importance that laws and regulations must be addressed and adapted for the technological arena. E-mails are generally broken into different packets that take different routes to reach their final destination. This is why, there are several points where the packets could be interpreted and examined. Without the presence of ethics and laws, a private e-mail message can become public. Piracy is possible over internet. It is very easy to duplicate an online application and send it over the internet.

Thousands of electronic bulletin boards that are present on

internet are used as clearing houses for transferring illegally produced copies of softwares that are copyrighted. The law is not clearly spelled out to determine what is reasonable and what is not. The problem is that it is easy to dispose of computer information. It is easy to send out information that might be incriminating to another location or person via internet. Since internet has made it possible for businesses to do business electronically, there is a rising concern that transactions can be intercepted as easily as email messages. With incredibly large sums of money exchanging hands over the internet, the potential for doing wrong is ever present. Communicating on internet is riskier as compared to communicating face-to-face in physical world because no one person is in charge of what happens on the internet. There is also great deal of concern regarding the accessibility of explicit material by children.

However, parents and organizations can block access and censor certain material and use systems that specify maturity levels to allow access to data. The wealth of information available on internet can easily be viewed, downloaded and passed on to other individuals. The right to use someone else's intellectual material should be granted by the owner of that information. They are protected by copyright laws as people who put them up on internet expect to receive credit for their work.

The use of internet has also facilitated some crimes in physical world occurrences. There is no doubt that crime will increase as technology grows because simpler access and ready availability have made internet a valuable tool for conventional criminals.

Internet is a tremendous resource and is a great place to learn, work and play. It draws its strengths from the diversity of the users because there is no one group or person in charge. However, with this continuous growth and change, it is important that it retains the qualities that have made it what it is today. For that, there must be ethics and laws with which people can operate.

Problems arise because people that cause these problems do not understand the full implications of their action.

7.5 Impact of Mobile Internet

Lia Puspitasari (2016) analyzed the role of mobile phones in narrowing the digital divide among Indonesian people at four levels—device ownership, internet adoption, use, and information acquisition.

Mobile phones can be used as a device for mobile leapfrogging (mobile leapfrogging refers to the process by which new internet users access the internet using mobile phones and not PCs). People access the mobile internet more for entertainment than for information gathering. The gap between developed and developing countries has reduced due to the use of internet. The digital divide occurs at the usage and skill levels as well as with the ownership of the devices. Mobile phones are valuable tools that allow leapfrogging because feature phone handsets are reasonable.

Demographic factors are considerably related to the use of the internet. Educational level is positively and significantly correlated with the adoption of the mobile internet on smart-phones but not on feature phones. Income is not significantly correlated with the adoption of using the mobile internet for either feature phones or smart-phones. Age is negatively correlated with the ownership of mobile phones, adoption of mobile internet, and mobile internet use, but not for feature phone ownership. Older people are less ready to adopt and use the internet even if they own mobile phones. The economic factors have a hindering effect on mobile network service consumption by the poor in spite of affordable handsets but it is not an important factor for narrowing higher levels of digital divide.

To sum up, internet has transformed lives of the people and has made people its slaves. One cannot imagine life without internet. Regarded as a powerful general purpose technology, internet quality is as important as internet access, if not more, as internet usage is not limited to e-mail and web browsing. With the interactive media-rich services and more user-generated contents being uploaded and shared, and with the availability of high quality videos, we cannot talk only about internet penetration.

8

VOICE Survey of Quality of Internet Services (QoIS) in India

As one hears the phrase *Industrial Revolution*, one can recollect the era in late 18th and 19th centuries, when rapid changes were taking place all over the world, led by England and America. Those were the changes which transformed the way manufacturing was done. This generation is witnessing another such evolution in the field of technology. The proliferation of internet is becoming a harbinger of new such type of industrial revolution inviting far reaching impact and changes in society which makes it imperative to assess the validity of this belief.

8.1 Four Phases of the Growth of Internet in India

In the 1990s, Department of Telecommunications (DoT) began investing in digital equipment. The government had realized that telecommunication services contribute to economic growth. The government then adopted ideas for reforms, with the objective of making India a hub of business opportunities. To achieve this objective, the government streamlined the foreign direct investment (FDI) process; worked to attract FDI in industries like telecommunications, software, and electronics; opened new investment avenues; created software technology parks; and identified new profitable sectors.

In India, leading ISPs like BSNL, Sify, Pronto Networks, and Dishnet Wireless have setup wi-fi networks at airports, coffee shops, and other locations in metropolitan areas. For instance, WiFyNet has installed a city-wide network in Mysore, which is the largest wi-fi hotspot in India. Intel and Pune municipality took an initiative to provide a city-wide wi-fi network.

We can view the growth of internet in India during four phases alongside the development of State Telecom Policy. They are National Telecom Policy, 1994, National Telecom Policy, 1999, Broadband Policy, 2004 and National Telecom Policy, 2012.

8.1.1 Phase I: Telecom Policy, 1994: At the onset of the first phase, the Government took upon itself to improving India's infrastructure, competitiveness, exports and its attractiveness to foreign investors through growth in telecommunications. Centrally-owned Videsh Sanchar Nigam Limited launched and operated internet services in India in 1995. In 1998, the government gave licenses to private operators to offer internet services in India. Failures and hindrances forced Government to revisit its policy framework which led to the second phase.

8.1.2 Phase II: National Telecom Policy, 1999: It was only in the second phase that the government recognized the relationship between telecommunications infrastructure development and social and economic growth of the country. When the intended results were not realized, government began to acknowledge the importance of telecom services. These were meant not only for the growth of the IT industry but for the development of the entire economy. For the first time in the history of Indian telecom, the government welcomed corporate involvement in providing paging services, cellular mobile telephone services (CMTS), and later, fixed telephone services (FTS). In addition, 100 percent FDI was allowed in providing internet services. These policies infused competitiveness in the information highway market and increased penetration of telecom services (Kaushik and Singh, 2004).

8.1.3 Phase III: Broadband Policy, 2004: The broadband policy was laid down by the government to realize the potential of broadband services. The policy was implemented with the objective of enhancing quality of life through high-speed access to information, web-based communication and better avenues of entertainment by enabling e-education, emedicine, e-governance etc. Prior to the Broadband Policy, broadband penetration in India was significantly low as compared to other Asian counterparts. At the time of implementation of the policy, the penetration of broadband, internet and personal computers stood at 0.02 percent, 0.4 percent and 8 percent respectively. There was no uniform standard for broadband speed and connectivity. Internet access speed varied from 64 kbps to 128 kbps. Following targets were set under the policy (Table 8.1).

Table 8.1 Targets of the Broadband Policy, 2004

(in millions)

Year ending	Internet subscribers	Broadband subscribers
2005	6	3
2007	18	9
2010	40	20

Source: cis-india.org.

The policy aimed at providing broadband (minimum speed of 256 kbps) to 20 million subscribers by 2010. However, there were only 13 million broadband subscribers as in May 2012. This led to the fourth phase of the growth of internet.

8.1.4 Phase IV: National Telecom Policy (NTP), 2012: NTP, 2012 envisioned *broadband on demand* and envisaged leveraging telecom infrastructure to enable all citizens and businesses, both in rural and urban areas, to participate in the internet and web economy, thereby ensuring equitable and inclusive development across the nation. The policy envisaged support to neutral services in e-governance and m-governance in key social sectors such as health, education and agriculture that were at that time limited to a few organizations in isolated pockets. This was expected to expand the footprint of these services and thus foster an atmosphere of participative democracy delivery model that was truly citizen-centric. Given the continued predominant role of wireless technologies in delivery of services in ICT sector, NTP, 2012, incorporated framework for increasing the availability of spectrum for telecom services including triple play services (voice, video and data) for which broadband was the key driver. This was to be facilitated by deployment of services through appropriate instrumentalities, while safeguarding the national interests.

8.2 Why is Quality of Internet Service (QoIS) Important?

Today, internet has become the lifeline of modern day businesses and helps common people in accomplishing their day-to-day tasks. Economies have progressed the world over and internet has been a major driver of this progress. Mankind has begun to rely on it to fulfil information and education needs, to say the least. Its depth and intensity to influence human life is way beyond what any of us could ever imagine. Presently, there are 7 billion people dwelling on this planet. We live in a society where everyone wants to stay connected and have their needs fulfilled in the shortest possible time.

Looking at some numbers to have a clear picture of how internet has become the most important innovation of this century and how it is impacting our society, we find that there are 3.2 billion internet users in the world, 207 billion emails are sent everyday through internet, 8.8 billion YouTube videos are watched everyday, 803 million tweets are sent in a day and 36 million is the number of Amazon purchases made in a day (ICRIER, 2016). Furthermore, according to Hostingfacts.com, there were 1.5 billion active Facebook users in the world, 2.9 billion Google searches are made every day, 2.7 million blog posts are published every day, 966 million is the number of websites in the world and these figures are growing exponentially.

The above facts provide evidence of majority of people's dependency on the internet in modern era. With time, this dependence is going to increase even more. About half of the world population is yet to be connected to the internet. The story is progressing almost on the same lines in India as well. According to a report by Statista, India is the second largest online market, with over 460 million internet users. However, it must be noted that only internet user figures give a partial picture as only 26 percent of the Indian population accessed the internet in 2015, despite the vast user base.

Not underestimating the need for *getting connected*, which is undoubtedly the essential first step, quality of services of internet has emerged as a critical variable. Regarded as a powerful general purpose technology, internet quality is as important as internet access, if not more, as internet usage is not limited to e-mail and web browsing. With the interactive media-rich services and more user-generated contents being uploaded and shared, and with the availability of high quality videos, we need to go much beyond internet penetration.

Since internet has been recognized as the driver of an economy's growth, it is important that this dynamic growth is accompanied by competition, efficiency as well as consumer satisfaction. This can only be possible if supported by a framework of well-structured regulations and clear roles of stakeholders. Thus, quality of service becomes an important parameter to gauge the performance of mobile internet services. Quality of service parameters ensures a healthy, competitive and transparent market. It is imperative for the consumers to be aware of and understand the actual terms of services offered. Since internet access is technical and intangible, it is difficult if not impossible for the consumers to evaluate QoIS by themselves. Also, since consumers lack technical know-how, they may not be able to determine the QoIS being delivered as compared to that being advertised by the service providers or the norms set by the regulators.

This VOICE study aims to assess the impact of quality of internet service on various aspects of life like education and research, business and jobs, availing health and medical services, development of rural areas, women empowerment, etc. through the perception of the users on the quality of internet service they are getting. Network quality in the internet services industry involves the quality and strength of the network signals, number of errors and, downloading and uploading speeds. Any break in the internet connectivity may lead to low perceptions of service quality. The quality of internet service is gauged on five salient parameters namely availability, speed, reliability, billing and charges and customer care services.

This study has evolved a theoretical framework to study the impact of quality of internet on users within the society. The framework has been developed mainly based on literature research. The footprint of the study is vast. Broadly, there are two sides of the framework—demand and supply—which are interactive in nature and share a give and take relationship. The supply side refers to the environment in which ISPs operate. This includes all ISPs, government policies, regulations and internet infrastructure enablers. The demand side is sub-divided into individuals and businesses. The demand side also refers to impacts on formal organizations such as schools, government agencies and private organizations. In very simple terms, the relationship between supply and demand factors can be illustrated as shown in Figure 8.1.

Figure 8.1: Relationships between Demand and Supply

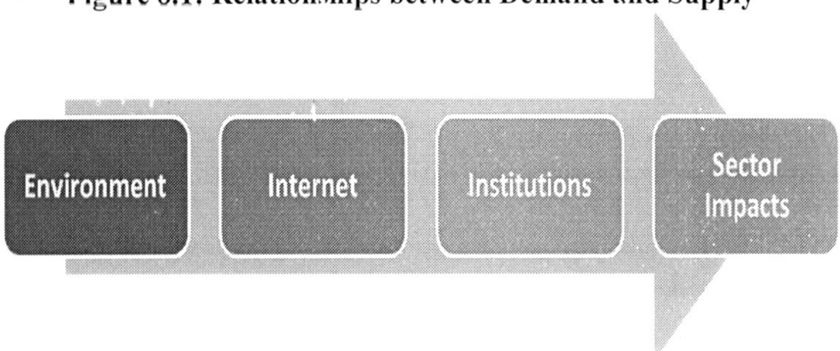

8.3 Objectives of the Study

VOICE has been conducting assessment of quality of service studies since 2005. The first study to assess the quality of telecom service, which covered mobile and broadband, was conducted in the year 2006-07. Since then VOICE has been regularly assessing QoS for the Telecom Regulatory Authority of India (TRAI). These studies involved assessment of quality of landline, cellular and broadband.

The purpose of the current study is to assess the impact of quality of internet service (QoIS) on consumers in realizing their day-to-day tasks and needs/goals and aspirations in their lives. There are various services for the consumers which are delivered through internet and consumers are accessing/using government and other services through internet. The study assesses the impact of quality of internet services delivered to its consumers and also list the shortcomings and impediments which require attention for further enhancing this quality.

Therefore, the broad objective of the current study can be enumerated as under:

1. To understand and enumerate different modes of accessing the internet.
2. To enumerate the profile of internet users w.r.t. their socio-economic background in urban and rural areas across the country.
3. To measure the perception of the customer about the quality of internet service from their service providers on access and use of internet through all modes.
4. To assess and enumerate the nature and consumer perception w.r.t. problems and grievance redressal mechanism in the usage of internet services.

The study also assesses the impact of internet and its quality of service on various aspects of life. Since the usage of any service is impacted by its role and purpose, the assessment of quality of internet will also require assessment of impact of internet and the quality thereof on various aspects of life.

8.4 Scope of the Study

The scope of this study was to capture consumer perceptions through various dimensions of life which are impacted by quality of internet service. The strategy of this study was to cover around 50,000 internet users in 19 States of India spread in all the telecom circles. TRAI has divided Indian states into four telecom circles. Metro circles cover Delhi, Mumbai and Kolkata; Circle A includes Gujarat,

Karnataka, Tamil Nadu, Andhra Pradesh including Telengana and Maharashtra including Goa; Circle B includes Haryana, Punjab, Kerala, Rajasthan, West Bengal, Uttar Pradesh (east), Uttar Pradesh (west) including Uttrakhand and Madhya Pradesh including Chhattisgarh; Circle C includes Bihar including Jharkhand, Odisha, Himachal Pradesh, Jammu & Kashmir, Assam and other North-eastern states of Arunachal Pradesh, Meghalaya, Mizoram, Nagaland, Manipur and Tripura. The study did not include the states of Jammu & Kashmir, Assam and North-eastern states due to some political, practical and logistical constraints.

In each State, the sample was further divided into rural as well as urban parts to get a holistic picture of the subject in study. The *internet user*, in context of this study, is anyone who meets a minimum level of knowledge about the uses of internet. It is presumed that not every internet user has sufficient knowledge of the uses of internet or how internet impacts various aspects of life. So, for the purpose of this study, internet user (or respondent) must have attained a minimum educational qualification of senior school certificate, to be able to understand the aspects under study and respond seriously. The respondents could be individuals or small businesses who had sufficient knowledge of their internet plans, and quality parameters like speed, reliability, billing and charges and help services.

The scope of the main study mainly subsumes three sub-studies:

1. **Perception on QoIS and its Impact on Users:** Capturing consumer perceptions on various parameters of quality of internet service (QoIS) that they are using and impact of QoIS on various aspects of life [1], captured through interviews with the respondents (face-to-face and CATI).
2. **E-governance:** Impact of quality of internet service while using e-governance services. Apart from capturing this aspect through main quantitative survey, some qualitative interviews were conducted with the providers and

beneficiaries of e-governance services. These interviews were videographed and a qualitative analysis was done thereon.

3. **Audit of Speed Test:** Internet speed is considered to be an important parameter of quality of internet service. Therefore, audit of wireless internet speed was also conducted in eight states in all the four telecom circles.

The entire study commenced in September 2016 and involved the following activities:

1. Literature review on internet, its impact on various dimensions and assessment as well as role of quality of internet service.
2. Methodology:
- Situation analysis through interaction with various stakeholders of telecom and internet service sector.
- Questionnaire building process for capturing the response of consumers, e-governance providers and beneficiaries and audit of internet speed testing.
- Pre-pilot test by research team and pilot test of questionnaires by field in eight states, analysis of data and changes in the questionnaire.
- Main survey in all the 19 States.
3. Data analysis and report writing.

Final report is based on a sample of 52,000 respondents.

8.5 Situational Analysis

Situational analysis involved in depth interviews with internet users (urban as well as rural), internet service providers (ISPs) and experts.

Experts interviewed included TRAI, Pehle India Foundation, National Institute of Public Finance and Policy, Gram Vaani, Indian Council for Research on International Economic Relations and others.

During situational analysis, more items were added to the list prepared post-literature review which came from in depth interviews with the respondents. The 199 items identified were increased to 225 items after the situational analysis. These factors were then put into importance-influence matrix (IIM). The factors were placed into different quadrants of IIM through rigorous brainstorming sessions. Following were the nine quadrants of the matrix.

Quadrant 1- Most important, most influential
Quadrant 2- Most important, influential
Quadrant 3- Most important, no influence
Quadrant 4- Important, most influential
Quadrant 5- Important, influential
Quadrant 6- Important, no influence
Quadrant 7- Not Important, most influential
Quadrant 8- Not important, influential
Quadrant 9- Not important, no influence

After necessary weeding out, 142 factors were retained.

Based on the 142 items retained, a first version of questionnaire was prepared. The questionnaire, thus prepared was put to test in the field in two phases: pre-pilot and pilot. Pre-pilot, conducted by the VOICE core research team, involved face-to-face interviews with 76 respondents from Delhi. It was primarily carried out to see what aspects can be merged with other aspects, whether the language of the questions was clear to the respondents, and to validate the sequence of the questions. The questionnaire was revised thereafter. This revised questionnaire was tested in the pilot survey which followed to capture the regional flavour besides testing the vitals of the questionnaire. The fact that India is a vast country with diverse language and culture, therefore, in order to assess the smooth flow and need of translating the questionnaire, pilot testing of the questionnaire was conducted in two states selected from each of the four circles.

The pilot test was conducted through face to face interviews and computer assisted telephonic interviews (CATI). A total of 1,046 interviews were conducted during the pilot which was about 2 percent of the total sample for the study. Post-pilot testing, a few changes were made in the questionnaire which included change of sequence of the factors in the questionnaire, for improved flow and better understanding and merging of some factors.

The final questionnaire thus obtained was then used for the main survey which started in the first week of May 2017 and concluded in August 2017.

8.6 Sampling and Data Collection

The total sample to be covered was 50,000, which was divided into 48,500 face-to-face using tablets and CATI (questionnaire-based) quantitative interviews. There was another sample of 2,400 for speed test audit sample using SIM based devices, and 120 in-depth (qualitative) interviews using audio visual facilities.

For this report, a total of 52,000 respondents were covered from all the telecom circles taken together. This sample comprised of 35,503 wireless and 16,497 wireline respondents spread over 3 metro cities (Delhi, Mumbai and Kolkata) and 15 States. 5 States were clubbed with neighbouring states. Thus, Uttar Pradesh included Uttarakhand, Madhya Pradesh included Chattisgarh, Maharashtra included Goa, Andhra Pradesh included Telengana and Tamil Nadu included Pondicherry. This means that the sample covered 3 metros and 20 States and Union Territories which should be considered a nationally representative sample. 80 percent of this sample was covered from the urban areas, leaving the rest 20 percent, covering rural counterparts, giving the sample a sufficiently large urban-rural comparison.

Endnote

1. This book covers only consumer perception on quality of service. Impact of quality of service on various aspects is not included in this book.

9

Time and Money Spent on Main Mode of Internet

Profiling of respondents in terms of the amount of time and money they spend on use of internet is an important aspect for studying the quality of service and gauging its impact. The questionnaire used for this study included questions that asked for active hours of usage in a day on various modes of accessing internet and average expenditure per month incurred on each of the modes on which the respondent uses internet. This chapter analyses the distribution of respondents on the basis of the time and money they spend on their main mode of using internet. This main mode is decided by the highest amount of time spent or by the preference of the respondent.

Money is a crucial factor in determination of choice of service provider and the plans they have to offer, whereas time is a measure of usage of service. These two factors together form important criteria in formation of impression of quality of service.

However, since the study was limited to the assessment of impact on consumers, following thresholds were set for these two factors for the purpose of the questionnaire based on pilot testing:

- For time, upper limit was set at 12 hours in a day for respondents using any mode.
- For money, due to the infrastructural difference between the modes, the upper limit for wireless respondents was ₹ 2,000; for wireline it was ₹ 3,500. The lower limit was fixed only for wireline respondents at ₹ 250 and not for wireless respondents as various monthly, weekly and even daily data packs are available.

9.1 Time Spent on Internet: Wireless Service

We now discuss the active hours of use in a day and expenditure per month of 35,503 respondents whose main mode of internet usage was reported to be wireless.

9.1.1 Telecom Circle-wise Time Spent on Wireless Internet:

- The mean time spent in a day, by respondents that claimed wireless to be their main mode of using internet was 3.5 hours. Mean time spent on internet was reported highest in circle C (4.2 hours).
- Most of the respondents in circle A and circle C had spent double (4 hours) the time spent by the respondents in metro circle (2 hours as reflected by mode).
- At least 25 percent (percentile 25) of the respondents from circle C spent less or equal to 3 hours in a day, whereas at least 75 percent (percentile 75) spent 5 hours or less in a day. This shows high usage of internet by respondents of circle C. One reason for this could be low speed of internet experienced by the respondents of circle C, which has been discussed later in this report.

9.1.2 Area-wise (Urban-Rural) Time Spent on Wireless Internet:

- The average mean time spent in a day by all wireless respondents was 3.5 hours. This was found to be slightly lower for rural respondents at 3.2 hours as compared to their urban counterparts (3.6 hours).
- However, most of the rural respondents spent 2 hours on internet while most of the urban respondents spent 4 hours, implying that most of the urban respondents were spending double the time on internet than the rural respondents.
- It can be said that at least 25 percent respondents living in both the areas spent 2 hours or less in a day and at least 75 percent of the sample spent 4 hours or less in a day.

9.1.3 Demography-wise Time Spent on Wireless Internet:

- Overall, the average mean time spent by wireless

respondents on internet was found to be 3.5 hours. Age-wise analysis shows that respondents aged between 19-25 years were found to be spending 3.6 hours. Also on an average, male respondents (3.5 hours) spent more time than females (3.3 hours). Unemployed and retired people were also spending more time (3.9 hours) in a day which is higher than the overall average usage time. *Others* which included mostly teachers, sales persons, clerks and managers were found to be spending much higher average of 4.5 hours on internet in a day.

- Most of the respondents that were aged above 36 years spent around 4 hours in a day on internet. It was also found that most of the general graduates and post-graduates spent double (4 hours) the time than most of the professionally qualified respondents and less educated HSC/SSC pass and diploma holders.
- At least 25 percent of the students spent up to 2 hours, but at least 75 percent spent at maximum 5 hours in a day. The same was the case with retired and unemployed as well.

9.1.4. SEC-wise Time Spent on Wireless Internet:

- Average time in a day spent on internet was found to be highest in the case of non-affluent lower class (SEC D and E) at 3.9 hours a day as compared to the middle (SEC B and C) and affluent upper class (SEC A). This is also reflected on by the percentile 75, where at least 75 percent of respondents belonging to lower classes (SEC D and E) spent 5 hours a day at maximum on internet as compared to 4 hours by other classes. This shows that lower the socio-economic class, more will be the time spent on wireless internet.

9.2 Money Spent on Internet: Wireless Service

9.2.1 Telecom Circle-wise Money Spent on Wireless Internet;

- The average mean expenditure per month of respondents covered during the survey on their main mode of using internet (wireless) was at ₹ 272. This average expenditure was found be higher in the case of circle A (₹ 280) followed by metro circle (₹ 278) and circle B (₹ 273). However, it was lower in the case of circle C at ₹ 228. The distribution of the sample into different categories of expenditure shows that:

- Half of the respondents spent between ₹ 250-₹ 499 (50.6 percent) on their wireless mode of accessing the internet. About 43 percent spent less than ₹ 249, whereas only 6 percent spent more than ₹ 500. However, in the case of circle C, over 62 percent spent less than ₹ 249, showing that the most of the respondents in circle C spent least amount using the internet. This is mainly because most respondents in this circle are from poor states (Bihar and Odisha).

9.2.2 Area-wise (Urban-Rural) Money Spent on Wireless Internet;

- Comparing the average expenses on internet of urban and rural population, it can be seen that urban respondents spent (₹ 282) more than rural respondents (₹ 246). When the distribution of respondents is seen through different categories, it can be ascertained that the percentage of rural respondents spending less than ₹ 250 is greater (60 percent) than that of urban respondents (41 percent).

9.2.3 Demography-wise Money Spent on Wireless Internet: The average expenditure of the respondents that claimed wireless as their main mode of using internet was ₹ 272 for a month. This average can be said to increase along the age categories with people aged between 14-25 years spending less than the average (₹ 202 and ₹ 248) and those above 25 years of age spending more than the average (₹ 283 and more). It can also be noticed that the amount spent on internet increased with completion of higher education levels as graduates (both general and professional) spent more than SSC/HSC pass and diploma holders.

The category-wise expenditure analysis of these respondents suggests that:

- The percentage of people spending less than ₹ 250 per month decreases with increasing age. Whereas, percentage of people spending more than ₹ 250 rises with increasing age.
- Every second male (49 percent) spends between ₹ 250-₹ 499, whereas this trend was found to be higher in the case of women (60 percent) respondents.
- More than three-fifth (62.9 percent) of the SSC/HSC pass respondents spent less than ₹ 250, but around half of graduates/post-graduates spent between ₹ 250-₹ 499.
- More than half of the respondents engaged in Government/public service, private service and business claimed to spend between ₹ 250-₹ 499. This percentage was 60 percent for housewives.

9.2.4 SEC-wise Money Spent on Wireless Internet: For all the wireless respondents, the average expenditure was ₹ 272 per month. It can be said that the amount spent on internet rises as we climb up the socio-economic ladder going from a low of ₹ 254 by the non-affluent lower class SEC E to ₹ 266 by middle class SEC B and C to ₹ 275 by the affluent class SEC A. The distribution of respondents into various categories shows that:

- Majority of the respondents from all SEC (around 50 percent) spent ₹ 250-₹ 499 on their monthly internet services.
- The proportion of people spending more than ₹ 500 on their wireless internet is highest in the case of affluent class SEC A (6.3 percent).
- The proportion of respondents spending less than ₹ 250 is highest for those belonging to lower class SEC D, where almost half of the respondents spent in this range (48.8 percent).

9.2.5: ISP-wise Money Spent on Wireless Internet: ISP-wise analysis shows that the highest average monthly expenditure was reported by Airtel users at ₹ 301. Contrary to the expectations, Uninor (₹ 168), Tata (₹ 250), Aircel (₹ 220) and Rel Comm (₹ 231) were cheaper than Reliance Jio as its users claimed to spend ₹ 253. The distribution of these users into various expenditure categories leads us to the following conclusions:

- Every one out of two (50.6 percent) respondents usually spendt between ₹ 250-₹ 499 on their internet service and 35.3 percent spent between ₹ 100-₹ 249. However, only 6 percent respondents remarked to have spent above ₹ 500.
- The highest percentage of people spending between ₹ 250-₹ 499 happened to be users of Reliance Jio (59.4 percent), closely followed by Airtel (53.9 percent).
- The percentage of Uninor users spending less than ₹ 250 was the highest among all other users (more than 80 percent).

9.3 Time Spent on Internet: Wireline Service

9.3.1 Telecom Circle-wise Time Spent on Wireline Internet:

- On an average, all respondents using wireline connection spent 4.3 hours on the internet. Respondents from metro circle recorded a mean of 3.8 hours—the lowest average among all telecom circles. However, the average time spent in circle B and circle C at 4.5 hrs and 5.4 hours respectively, was higher than sample mean (overall average).
- Most of the respondents in each circle individually spent 4 hours on internet, except metro circle where they spent half of this time (2 hours).
- It can be said that, at least 75 percent of the respondents belonging to circle B and C spent 6 hours or less in a day on internet. High use of internet in circle C is also confirmed by percentile 25, showing that at least 25 percent of the respondents spent up to 4 hours on internet as compared to 3 hours or less in other circles. As discussed earlier, low speed of internet connection is inversely proportional to high amount of time. In other words, a consumer with high speed connection ends up

downloading more data in less amount of time.

9.3.2 Area-wise (Urban-Rural) Time Spent on Wireline Internet:

- Area-wise analysis shows that urban respondents have reported spending more than the average time at 4.4 hours a day.
- The percentile 25 and percentile 75 figures show that in urban as well as rural areas at least 25 percent spent up to 3 hours on internet and at least 75 percent also spent around 6 hours or less.

9.3.3 Demography-wise Time Spent on Wireline Internet:

- Age-wise analysis shows that youngest lot aged between 15-18 years and those above 46 years spent less than average time (4.3 hours) on internet.
- On an average, males spent 4.4 hours as compared to females that spent 3.9 hours. However, for both males and females, it can be said that at least 50 percent spent between 3-5 hours a day on internet.
- It can also be observed that with the rise in educational qualifications, the average time spent on internet increases substantially (from 3.9 to 4.5 hours). Most of the highly educated respondents, i.e. graduates and post-graduates (general and professional) spent more time than spent by most of less educated HSC/SSC pass and diploma holders (mode).
- Profession-wise, most of the respondents that were engaged in government service spent 5 hours a day on internet use. Whereas most of the housewives reported to have spent only 3 hours.

9.3.4 SEC-wise Time Spent on Wireline Internet:

- Most (mode value) of the respondents from the entire SEC categories had reported that they spent 4 hours on internet. However, the mean value shows that affluent class (SEC A) spent less time (4.2 hours) than the non-affluent middle class, SEC B and C (4.5-4.6) and lower class (4.7-4.8).

This could be on account of type of speed plan (2G/3G/4G) subscribed by the affluent and non-affluent classes.

- At least 75 percent of the respondents belonging to SEC B, C, D and E can be said to have spent 6 hours or less in day on their wired internet service.

9.4 Money Spent on Internet: Wireline Service

9.4.1 Telecom Circle-wise Money Spent on the basis of Telecom Circle: Overall, the average expenditure of the respondents surveyed that claimed wireline to be their main mode of internet use was ₹ 895. The highest average expenditure was found in circle A at ₹ 991 and the lowest for circle C at ₹ 760. Respondents from metro (₹ 848) and circle B (₹ 837) were also spending less than the overall average. Distribution of these respondents into internet expense categories shows that:

- More than half of the respondents (55.4 percent) spent between ₹ 500-₹ 999 on wireline access to internet. This was reported highest in circle C (62.2 percent), followed by those in circle B (58 percent).
- Largest percentage of people spending between ₹ 250-₹ 499 was that in circle C (19.3 percent). However, overall only 9.6 percent of all wireline respondents spent in this range.
- Around 35 percent of the respondents spent above ₹ 1,000. This was reported highest by the respondents in circle A (43 percent) and lowest in circle C (18.5 percent).

9.4.2 Area-wise (Urban-Rural) Money Spent on Wireline Internet: Area-wise analysis showed that urban respondents tend to spend much more (₹ 900) than the rural respondents (₹ 778). The analysis of spread of the sample into various categories of expenditure indicated that:

- While 55.4 percent respondents reported spending between ₹ 500-₹ 999, about one in ten claimed to spend lesser between ₹ 251-₹ 500 (9.6 percent). About one-fourth (24.5 percent) reported spending more between ₹ 1,001-₹ 1,500.

The rest 10.5 percent reportedly spent above ₹ 1,500.

- For the two expenditure categories up to ₹ 1,000, the percentage of rural respondents was comparatively higher than that of urban respondents. However, as we go beyond ₹ 1,000, the order reverses.

9.4.3 Demography-wise Money Spent on the Wireline Internet: The average expenditure of all wireline respondents for a month was ₹ 895. Amongst all age categories, it was the teenagers aged between 15-18 years that spent lowest at ₹ 767 a month. Respondents aged above 26 years had spent much more than the average with those between 46-55 years spending ₹ 943 per month. It can be also observed that with rising level of education, the average expenditure increases.

The distribution of respondents on the basis of their expenditure on wireline connection hints the following impressions:

- More than half (55.4 percent) of the wireline respondents had their expenditure ranging between ₹ 500-₹ 999. Another quarter (24.5 percent) of them spends more than ₹ 1,000 a month.
- The percentage of people spending ₹ 999 and below decreased with increasing age, whereas for those spending between ₹ 1,001-₹ 1,500, it was vice-versa.
- For all expenditure categories, the proportion of males and females was more or less same.
- Every seven in ten housewives (67.8 percent) claimed to spend between ₹ 501-₹ 1000 on internet. Whereas, 60 percent students spent in this range. Among those who spent more than ₹ 1,499 a month, people engaged in private service and business, had the highest proportion.

9.4.4 SEC-wise Money Spent on Wireline Internet: SEC-wise analysis reveals that the middle classes, SEC B and SEC C, spent more than the average—₹ 898 and ₹ 927 respectively. This average went decreasing with the reducing economic capacities of lower classes and came down to as low as ₹ 760 for SEC E. The class-wise analysis of distribution of respondents into different categories of expenditure yields the following points:

- More than half of the respondents (55.4 percent) had spent between ₹ 500-₹ 999 on their wired access to internet. The largest fraction of such people was affluent class, SEC A (55.9 percent), closely followed by middle class SEC B (55.4 percent) and lowest in lower class SEC D (46.4 percent).

9.4.5 ISP-wise Money Spent on Wireline Internet: ISP-wise analysis suggests that the average expenditure by users of You Broadband was highest at ₹ 1,333, closely followed by those of Hathway at ₹ 1,011. However, for other regional providers like Asianet and Connect, the average expenses were much cheaper at ₹ 690 and ₹ 782 respectively. The scrutiny of division of respondents into various expenditure categories reveals as follows:

- More than half of the respondents (55.4 percent) spent between ₹ 500-₹ 999 on their wireline internet connection. However, this proportion was highest for users of Tikona and Connect (more than 75 percent) and least for that of You Broadband users (18.6 percent).
- About 10 percent respondents claimed to have spent between ₹ 250-₹ 499. The lowest such section of users were that of You Broadband (0.8 percent) and Tikona (1.7 percent).
- Amongst the 24.5 percent respondents spending between ₹ 1,000-₹ 1499, the users of GTPL were highest in percentage (34.2 percent).

9.5. Relation between Time and Money

When the mean values of time and money are placed in contrast, it leads to the following interesting insights:

- The weakest telecom circle in terms of user density and infrastructure, circle C had the highest average hours of usage on both modes of service—wireless (4.2 hours) and wireline (5.4 hours). However, the mean expenditure on both modes is lowest for the same circle—wireless (₹ 228) and wireline (₹ 760).

- Respondents living in rural areas spend almost the same time as that of their urban counterparts on wireline at 4.3 hours in spite of wireline being the costlier mode of internet access.
- SEC D and E from the less affluent lower class spent the highest time among all classes on wireless (3.9 hours) and wireline (4.7 and 4.8 hours) respectively even when the expenditure incurred by them is lesser.

Any service is priced on basis of its offerings, thus, the money spent on internet service informs about the download and upload speeds and the data allowance that has been purchased. Given the fact that data consumption is a function of speed, it can be deduced that higher speeds cause increased used of data at the same time as compared to that of low speed. Therefore, for all the three findings mentioned above, it can be safely assumed that the purchased services were not as good as that of the others in the same category. It is due to this that respondents in circle C, rural areas and lower classes spend relatively more time on an average on their main mode of internet use.

These arguments and observations thus lead us to the conclusion that quality of internet service which is defined as an interplay of various parameters on the grounds of time spent on internet use, in return, leads to the estimation of time that a user will take for performing an activity through internet that is purchased after spending some money. It is like a vicious circle. But, nevertheless delivering promised quality of service stays imperative for efficient use of time and money.

10

Quality of Internet Services

10.1 Quality of Service (QoS) Defined

The International Telecom Union defines quality of service (QoS) as the 'totality of characteristics of a telecommunication service that bears on its ability to satisfy stated and implied needs of the user of the service'.

TRAI has revised this definition to corroborate all kinds of services that come under its umbrella, which says that 'QoS is the main indicator of the performance of a telecommunication network and the degree to which the network conforms to the stipulated norms. The subscriber's perception of the quality of service (QoS) is determined by a number of performance factors'.

10.2 Quality of Internet Services (QoIS)

QoIS, according to RF Benchmark, "specifies the guaranteed level of quality services ordered and received by the mobile network user during the service contract with the operator. As far as telecommunication services are concerned, QoIS is defined by three functions: access, information transfer and disconnection, and three criteria: speed, accuracy and reliability. The mobile network operator is obliged to provide services of a specific, guaranteed quality".

Until some years ago, broadband gap used to mean how many people have access to internet as against those who do not have access to internet. However, broadband divide is no longer viewed as a penetration divide. It is now becoming more of a quality and capacity divide. When talking about broadband divide, "the relevant questions should be how good the connection is? How fast it is? And how fast is fast? It is therefore, a divide in the range of services people can access and use. Governments, policy makers and regulators, around the world, have made increasing the internet penetration their main priority". Not underestimating the need for *getting connected*, which is undoubtedly the essential first step, quality of services cannot be ignored (Vicente and Gil-de-Bernabé, "Assessing the Broadband Gap: From the Penetration Divide to the Quality Divide", 2010).

Since internet has been recognized as the driver of an economy's growth, it is important that this dynamic growth is accompanied by competition (demonstrated by consumer choice), efficiency as well as consumer satisfaction. The significance of competition can be explained with the help of an example. Suppose, in an area in Delhi, only one service provider offers its services; no other service provider serves in that area.

Despite people knowing that the service provider X provides better services, they cannot avail of its services because it is not available in that particular area. They are forced to avail of the services offered by that 'only' service provider available in that area, howsoever poor it may be. This is illustrated by local monopolies in parts of NCR Delhi where only one ISP is available.

Healthy competition is thus necessary for the market to function efficiently. This can only be possible if supported by a framework of well-structured regulations and clear roles of stakeholders. Quality of service, thus, becomes an important parameter to gauge the performance of mobile internet services. Quality of service parameters ensures a healthy, competitive and transparent market but for that, it is imperative for the consumers to be aware of the actual terms of services offered. Since internet access is technical and intangible, it is difficult if not impossible for the consumers to evaluate the QoS by themselves. Also, since consumers lack technical know-how, they may not be able to determine the QoS being delivered as compared to that being advertised by the service providers or the norms set by the regulators (CUTS, 2016).

10.3 Regulation of QoIS

TRAI regulates the QoIS and monitors the performance of internet service providers (ISPs) in India. TRAI contrived standards for mobile internet under the QoS for Wireless Data Services Regulations, 2012 in which TRAI laid down stringent quality guidelines for wireless data services that telecom operators have to abide by. The regulations mandate service providers rendering unified access service (UAS) as well as cellular mobile telephone service to submit monthly performance reports to TRAI. The regulator based on these reports, releases quarterly report for all the operators in the 22 service areas in India. However, no penalties are imposed on operators that do not abide by these quality guidelines and likewise, there are no incentives (from the regulator) for those performing well.

In July 2014, TRAI introduced a new clause that mandates service providers to state minimum download speed for the wireless data plans. The clause requires operators to ensure that the download speed advertised is delivered not less than 80 percent of the usage time. This clause was introduced as the advertised speeds differed significantly with what was being delivered. TRAI has separate QoS regulations for broadband and wireless data services. 3G and 4G data services can be categorized as broadband services but 2G mobile internet services cannot be regarded as broadband service. Given the fact that a significantly large chunk of population still uses 2G, there is a separate set of regulations required for wireless internet services, at least, till the time the users choose to upgrade to faster internet services (CUTS, 2016).

TRAI issued a mandate stating that all fixed line internet providers will always provide a minimum speed of 512 kbps to their users in order to be classified as a broadband service provider. This order includes all the services that offer a fair usage policy (FUP). Under FUP, an internet service has to provide an internet connection to the user at all times at a certain minimum guaranteed speed. So, instead of the earlier

64 kbps download speed, users will now get a minimum download speed of 512 kbps after they have used up their data at higher speeds. TRAI has also asked all internet providers, both wired and wireless, to ensure that the full information about their FUP plans is accessible to the public (Mehta, The Huffington Post, 2016).

In an attempt to improve internet service quality for Indian mobile users, TRAI launched an application TRAI MySpeed App on July 5, 2016, to help consumers check real time internet speed on their hand held devices which they can report to the regulator. The data, thus collected from the subscribers, was expected to be used by TRAI to fix new quality of service norms. Most of this data is hidden and only a general average is displayed online.

As far as the quality parameters are concerned, speed is an important determinant of internet quality. However, there are additional factors that come into play, when determining the quality of internet services. Broadband experience is affected by throughput, latency, network oversubscription, jitter and packet loss etc. However, it must be noted that for consumers, the only available information about quality is typically download speed. There are factors beyond technical performance of broadband that affect the overall quality of experience. These may include cost, ease-of-use, customer support and user expectations. Also, significance of various quality parameters differs across mobile applications. Some need low latency while others may require good network reliability, and yet others need great bandwidth and speed and so on. In order to create a mass market for innovative next generation applications, networks need to consistently support high bit rates in both the directions—low latency and reliable connectivity.

Thus, a country cannot properly assess its progress on digitization without taking into account the service quality.

In order to fill this void, the industry regulator—Telecom Regulatory Authority of India (TRAI)—has been mandated under TRAI Act, 1997 to lay down quality of service (QoS), to ensure QoS and to conduct periodical surveys to assess QoS so as to protect the interests of the consumers. The authority has been exercising these provisions that are linked with the objective of protecting interests of service providers and consumers and maintaining an environment of orderly growth in the industry. However, it has stopped regular QoS surveys since 2015.

Over the years, TRAI has released a number of regulations for basic wireline, cellular mobile, broadband, dial-up and leased line connections and wireless services. These regulations have been reviewed time and again as evolving customer requirements and expectations have been given paramount considerations. The framing and moderation of such regulations is the result of a consultation process through written comments and open-house discussions, which comprehensively addressed all aspects of QoS. Such discussions with service providers, consumer organizations and general public are held in various parts of the country to get inputs from the stakeholders. The comments and feedback received from stakeholders are thus taken into account.

The regulations clearly define the accountability of the service providers towards the authority, failing which they could be either penalized in the form of financial disincentives or get their licenses revoked. In addition to prescribing the benchmarks for various parameters of QoS with their denotation, the authority also specifies the measurement methodology, record-keeping, review and audit requirements to be adhered by the service provider. It also reserves the right to publish such reports on regular basis to ensure that consumers have adequate information about the QoS being provided by various service providers.

These results are released in the form of performance indicator reports at the end of each quarter and available at the website (Telecom Regulatory Authority of India, 2016) for perusal. Furthermore, TRAI has been conducting periodic surveys through independent agencies across the country, to monitor the compliance of prescribed standards/benchmarks. Zone-wise tenders were floated on an annual basis wherein the research agency had to present the findings in a tabular format that depicts which service provider has not met the quality of service benchmarks of the prescribed parameters. These regular survey reports have stopped being published in 2015 and have not been resumed since then.

10.4 Survey on QoIS

The total number of respondents covered during the fieldwork was 52,000, out of which 35,503 (68.3 percent) were those who reported wireless as their main mode of internet use and 16,497 (31.7 percent) reported wireline as their main mode of internet use. These respondents were users of 12 wireless ISPs or 92 wireline ISPs. On the basis of main mode of internet use, i.e. wireless and wireline, for the total sample and for the sample of each telecom circle individually, the ISPs that had greater than 1 percent of the sample size and were more than 30 in count have been taken into consideration for analytical comparison and ranking as they had significant sample sizes. All the ISPs that did not meet these two conditions have been clubbed into 'others'. The only two public sector ISPs— MTNL (serving in Delhi and Mumbai only) and BSNL (serving all-India except Delhi and Mumbai) have been taken together for the purpose of analysis.

10.5 Parameters of Quality of Service

In order to assess the quality of internet services, this study sought ratings from the respondents on a number of sub-parameters under following five constituent parameters of quality on a scale of 0 to 5, where 1 meant poor and 5 meant excellent and 0 stands for 'can't say'.

1. Availability.
2. Speed.
3. Reliability.

4. Billing and charges.
5. Help and customer care services.

A total of 18 sub-parameters were classified under these five major categories as question no. 5 in the questionnaire. However, in case of respondents whose main mode of internet was wireless, did not rate on *'provision of technicians as and when required'* under parameter—'help and customer care services'; and in case of respondents whose main mode of internet was wireline, did not rate on *'uninterrupted services at all places'* and *'effective roaming services'* under parameter—reliability, as these were not applicable in their respective cases. The mean values for the five main parameters have been derived through the computation of mean of all sub-parameters, excluding all the 'can't say' responses. Other than these five parameters of quality, respondents were also asked to rate 'overall quality of their internet service' on the scale of 1-5, where 1 was poor and 5 was excellent.

10.5.1 All Circles:

A. Wireless: Out of the 52,000 respondents covered during the survey, 35,503 (68.3 percent) respondents were those who reported using wireless i.e. mobile or dongle as their main mode of internet.

Amongst the parameters, the mean rating of all ISPs taken together on 'overall availability' (4.23) was highest, followed by 'overall help and customer care services' (4.22) and 'overall speed' (4.21). The lowest average rating of 4.20 was on two of the important parameters—overall billing and charges and overall reliability. This observation can be explained through literature that justifies that getting information about a connection and its services is the easiest but the assurance that the connection will work seamlessly without hiccups is really hard to get.

ISP-wise comparisons showed that:

- The brand new ISP—Reliance Jio held top spots on all five parameters of quality.
- Two National ISPs—Airtel and Rel Comm have thus been

rated on second and third slot respectively on all parameters except 'overall availability' where Rel Comm (4.29) was rated higher than that of Airtel (4.27). All these ratings were more than the average.

- However, Idea was rated marginally below average on all parameters.
- On the lower end were Uninor and Aircel that were rated least on each parameter in the same order. Other national ISPs like Tata, MTNL/BSNL and Vodafone occupied mid positions on the rating scale with ratings quite less than the overall averages.

B. Wireline: Altogether 16,497 (31.7 percent) respondents out of 52,000 covered during the entire fieldwork had reported using wireline as their main mode of internet. A number of regional and local wireline ISPs were clubbed into 'others' category for they did not have a significant sample size and thus could not be compared individually.

Amongst the parameters, 'overall reliability' had the highest rating of 4.31, supporting the theory that fixed connections are more reliable. This was followed by 'overall availability' (4.29) and the third highest rating was that of 'overall help/customer care services' at 4.26. 'Overall billing and charges' and 'overall speed' were rated lowest at 4.25 each.

ISP-wise analysis suggests that:

- ISPs with national license—You Broadband and Hathway were rated highest and second-highest respectively on all parameters with low variations in their ratings. Other national ISPs like Rel Comm, Tata and MTNL/BSNL had marginally less than average ratings.
- Ratings achieved on all five parameters of quality by GTPL (serving in Gujarat, Maharashtra and Rajasthan), Tikona and Airtel were all above average. Similarly, Alliance Broadband (serving in West Bengal) was also rated above average on all parameters but 'overall reliability' where its rating was 4.18 i.e. quite below the average of 4.31.

- Other national wireline SPs like Tata and Rel Comm were rated less than average on all parameters. This was also the case for the only public sector service providers available in the country—MTNL/BSNL.
- Connect Broadband (serving in Punjab, Haryana and Himachal Pradesh) were rated lowest on each parameter.
- It is pertinent to point that national ISP Airtel, which has largest subscriber base has lower rating than Hathway and You Broadband on all five parameters.

10.5.2 Metro Circle:

A. Wireless: Altogether 5,172 respondents were covered in the metro circle, of which 3,422 (66.2 percent) reported wireless as their main mode of internet service and rated it on various parameters of quality.

When we see parameter-wise, the average rating of 'overall availability' was highest at 4.15, followed jointly by 'overall help and customer care services' and 'overall speed' at 4.12 each. The third highest rating was of 'overall billing and charges' at 4.11 and the lowest was thus of 'overall reliability' at 4.09.

When we compared these mean ratings according to ISP, we can observe that:

- Tata (4.36) was rated quite higher than Reliance Jio (4.24) on the scale of 'overall availability'. The ratings of Airtel (4.26) and MTNL/BSNL (4.15) were also found to be above average.
- On all parameters of quality, the ratings achieved by Reliance Jio were above or equal to that of Airtel, however the mean difference was significant <0.05 in each case.
- Apart from 'overall availability', Tata and MTNL/BSNL were rated above average on all parameters except 'overall speed' and 'overall help/customer care services' respectively.
- Aircel were found to be on the lower end of the rating scales for all five parameters except 'overall help/customer care services', where the lowest rating was that of Rel

Comm.

- Two other National SPs—Idea and Vodafone were rated below average on each parameter, except 'overall speed' where the rating of Vodafone was equal to average.

B. Wireline: Altogether, 1,750 (33.8 percent) respondents out of 5,172 respondents of the metro circle reported using wireline as their main mode of internet.

Amongst the parameters, the average rating of 'overall reliability' and 'overall availability' was highest at 4.25 each, followed by 'overall help/customer care services' at 4.23. the lowest was that of 'overall speed' and 'overall billing and charges' at 4.22 each.

Comparing these ISP-wise, it was observed that:

- Spectranet was rated most reliable ISP with rating of 4.58. With huge difference in these ratings, the second most reliable ISP was Rel Comm, Hathway, and Tikona with mean ratings of 4.45, 4.43 and 4.37 respectively.
- Tikona was rated highest, followed by Rel Comm on 'overall availability', 'overall speed' and 'overall help and customer care services'. However, Rel Comm toppled Tikona on parameter 'overall billing and charges' to gain the top position with almost equal mean ratings of 4.33 and 4.32 respectively.
- Hathway achieved above average ratings on all parameters of quality. Airtel was rated above average on all parameters, but not on 'overall reliability'. Similarly, Spectranet was not rated above average only on 'overall availability'.
- It was only MTNL/BSNL and Alliance Broadband (serving in West Bengal) that achieved less than average ratings on all five parameters, where the lowest ratings were that of the latter.

The most significant finding is that standalone ISPs like Spectranet, Hathway and Tikona were rated higher than telecom giants like Airtel, MTNL/BSNL and Rel Comm.

10.5.3 Circle A:

A. Wireless: In circle A, 20,886 respondents were covered of which 14,156 (67.8 percent) reported wireless as their main mode of internet service and rated it on various parameters of quality. It can be noted that the average rating of 'overall help and customer care services' was highest at 4.29. Two parameters namely 'overall availability' and 'overall speed' were rated second highest at 4.28. This was followed by 'overall billing and charges' (4.27) and 'overall reliability' (4.27), which were the lowest.

The ISP-wise analysis showed that:

- Reliance Jio was rated highest on all five parameters of quality.
- The second highest rating on all parameters was attained by Rel Comm except on overall speed, on which Airtel was the second highest rated ISP with a rating of 4.29.
- Airtel was rated above average on all parameters of quality except on 'overall billing and charges' where the rating of 4.26 was marginally lower than the average rating of 4.27.
- All other national ISPs—Aircel, Idea, Vodafone and BSNL were found to have below average ratings for all parameters of quality. Also, Tata was rated lowest on all parameters.

B. Wireline: Altogether 6,730 (32.2 percent) respondents out of 20,886 respondents of the circle A reported using wireline as their main mode of internet.

Amongst the parameters, the average rating of overall reliability was highest at 4.37. Second highest was 4.34 of 'overall availability', followed by 'overall help and customer care services' (4.33), 'overall billing and charges' with mean rating of 4.32. The lowest was that of overall speed at 4.31.

The ISP-wise analysis of these average ratings indicated that:

- Regional ISPs like Beamtele and Spidigo have outshone national ISPs like Airtel, Rel Comm, BSNL and Tata on all parameters of quality. Beamtele that serves in Andhra

Pradesh was rated highest on 'overall availability', 'overall help and customer care services' and 'overall speed' with ratings of 4.66, 4.61 and 4.63 respectively. Spidigo that serves in Gujarat and Maharashtra was rated highest on 'overall reliability' and 'overall billing and charges' with ratings of 4.90 and 4.66 respectively.

- The third, fourth and fifth highest ratings on all parameters were attained by national ISPs You Broadband, Hathway and Tikona in the same order.
- National ISPs—Airtel and Rel Comm and regional ISP-GTPL that serves in Gujarat, Maharashtra and Rajasthan also achieved above average ratings on at least three of the five parameters of quality.
- Tata was rated least on all the parameters. Apart from Tata, it was only public sector ISP—BSNL—that was rated below average on all parameters.

It shows that smaller regional players like Beamtele and Spidigo are able to provide better quality of service than giant telecom companies which have a national footprint.

10.5.4 Circle B:

A. Wireless: In circle B, 21,120 respondents were covered, of which 14,499 (68.7 percent) reported wireless as their main mode of internet service and rated it on various parameters of quality.

The highest average rating was attained by 'overall availability' at 4.25. Lower at 4.21, was 'overall help and customer care services' as the second highest rated parameter for all ISPs taken together. This was closely followed by 'overall speed' (4.20) and least rated were 'overall billing and charges' and 'overall reliability' at 4.19 each.

When the ratings were looked into ISP-wise, they showed that:

- For each of the five parameters of quality, the scores of Rel Comm were remarkably higher than that of the second highest i.e. Airtel. Quite near to the Airtel, the third highest ratings were obtained by Reliance Jio on all parameters.

- Idea was rated above average on all parameters except 'overall availability' where its rating of 4.24 was marginally less than average (4.25). BSNL was rated above average only on 'overall availability' with mean ratings of 4.27.
- Aircel, Tata and Vodafone were rated below average on all parameters of quality. The least ratings on 'overall availability', 'overall reliability' and 'overall billing and charges' were attained by Vodafone. Whereas, Tata was rated lowest on 'overall speed' and 'overall help/customer care services'.

B. Wireline: 6,621 (31.3 percent) respondents out of 21,120 respondents covered in circle B reported using wireline as their main mode of internet.

Amongst the parameters, the highest average rating was that of 'overall reliability' at 4.28, closely followed by 'overall availability' with rating of 4.26. On the lower end, rest three parameters—'overall speed', 'overall help and customer care services' and 'overall billing and charges' were least rated at 4.20, 4.20 and 4.19 respectively.

When a comparison of these average ratings was drawn on the basis of ISPs, it was noticed that:

- Airtel was the only ISP that was rated above average on 'overall reliability'. This mean rating of 4.44 was also the highest.
- Airtel and Alliance Broadband from West Bengal were top rated in the same order on 'overall availability', with mean ratings of 4.38 and 4.34 respectively. However, on the rest three parameters—'overall speed', 'overall billing and charges' and 'overall help/customer care services', the two switched positions.

Rel Comm, BSNL, Tikona and Asianet (serving in Kerala) were rated below average on all parameters. Whereas, Connect Broadband (serving in Himachal Pradesh, Haryana and Punjab) was least rated on all parameters.

10.5.5 Circle C:

A. Wireless: In circle C, 4,822 respondents were covered of which 3,426 (71 percent) reported wireless as their main mode of internet service and rated it on various parameters of quality. Interestingly, the findings of this circle were quite different from those of the others.

The highest average rating was attained on parameter 'overall help and customer care services' i.e. 4.07. This was slightly above 4.06 i.e. average rating on 'overall speed', 'overall billing and charges' and 'overall reliability'. The least average rating was that of 'overall availability' at 3.98, which is actually opposite to expectations.

The observations from ISP-wise analysis showed that:

- Airtel attained the highest ratings on all parameters of quality, thereby holding the top position.
- Reliance Jio was rated above average on parameters but that of 'overall billing and charges' where its rating of 4.05 was marginally less than the average rating (4.06).
- Two National ISPs—Vodafone and Uninor that were rated quite low in all other circles have been rated above average in circle C on all five parameters of quality.
- All other National ISPs—Aircel, Idea, Rel Comm, Tata and BSNL attained below average ratings on all parameters, where Tata was least rated.

The ratings in this circle were somewhat lower than other circles.

B. Wireline: 1,396 (29 percent) respondents out of 4,822 respondents of circle C reported wireline as their main mode of internet use. The ratings given by these respondents on four wireline ISPs on various parameters of quality have been presented in the form of mean ratings.

Amongst the parameters, the highest average score was that of overall reliability at 4.31. The descending order so followed had 'overall billing and charges' (4.27), 'overall speed' (4.25), overall help and customer care services' (4.24) and overall availability at 4.22, being the lowest one in sequence.

These ratings when observed in contrast to the ISP, reflected the following:

- Airtel was rated highest on all five parameters of quality. Apart from this, it was only Ortel that attained above average ratings on 'overall reliability', 'overall billing and charges' and 'overall 'help/customer care services'.
- National SPs BSNL and Rel Comm have been rated less than average on all parameters individually, whereas the latter was the lowest on all five parameters.

10.6 Overall Quality of Service

10.6.1 All circles:

A. Wireless: The average rating of all ISPs on overall QoIS was 4.14. However, Airtel and Reliance Jio were rated more than average with ratings at 4.25 and 4.18 respectively. Slightly less than this average rating was Idea (4.11). Other national SPs like Vodafone, Rel Comm, MTNL/BSNL, Tata and Aircel were rated lesser at 4.07, 4.07, 4.05, 4.04 and 4.01 respectively. Uninor was least rated with mean rating of 3.88.

B. Wireline: The average rating of all ISPs on overall QoIS was 4.18. You Broadband, Hathway, GTPL, Airtel and Tikona were rated above average at 4.71, 4.55, 4.46, 4.24 and 4.20 respectively. All other ISPs, namely Rel Comm, Asianet, MTNL/BSNL, Tata, Alliance and Connect were rated below average and lowest rating was 3.95 attained by Connect.

10.6.2 Metro Circle:

A. Wireless: The average rating of overall QoIS of all ISPs was 4.05. Reliance Jio (4.15), Airtel (4.12) and Tata (4.20) were rated substantially more than this average. Equal to average (4.05) was Vodafone's rating. All other ISPs were rated less with Rel Comm being the least rated SP at 3.76.

B. Wireline: The average rating of all ISPs on overall QoIS was 4.14. The top three rankers were rated quite higher than this average. They were Spectranet, Hathway and Tikona with ratings of 4.49, 4.42 and 4.37 respectively. Rel Comm

(4.17) was also rated above average. The least rating of 4.06 was obtained by MTNL/BSNL and Alliance jointly.

6.2.3. Circle A:

A. Wireless: The average rating of overall QoIS of all ISPs was 4.22. Reliance Jio was rated highest at 4.36. It was followed by Airtel which had a rating of 4.23. Only these two were rated above average. The third position was sought by Idea (4.16), followed by Vodafone with a rating of 4.14. Aircel with a rating of 4.00 was the least rated.

B. Wireline: The average rating of all ISPs on overall QoIS was 4.25. Two regional ISPs were extraordinarily high in their ratings of overall QoIS. They were Spidigo and Beamtele with scores of 4.93 and 4.91 respectively. On the third rank was You Broadband with quite a difference having a rating of 4.82. However, these providers have outshone all the other national ISPs like Hathway (4.61) and Tikona (4.41). GTPL was also rated above average at 4.46. Airtel was marginally below average at 4.24. The lowest rating of 4.06 was attained by Tata.

10.6.4 Circle B:

A. Wireless: The average rating of overall QoIS of all ISPs was 4.12. Quite different from what was happening in other circles, here, Rel Comm was rated highest at 4.21. It was followed by Airtel which had a rating of 4.18. The third position was shared by BSNL, Reliance Jio and Aircel with a mean rating of 4.13 each. Idea and Tata were rated below average. Barring other ISPs that had insignificant sample size, national ISP Vodafone was rated lowest at 4.03.

B. Wireline: The average rating of all ISPs on overall QoIS was 4.13. It can be noticed that only Airtel was rated more than this average, i.e. 4.27. Slightly lower than average was Asianet having a mean rating of 4.10. Amongst the lowest performers were Alliance, Tikona and Connect with mean ratings of 3.97, 3.96 and 3.95 respectively.

10.6.5. Circle C:

A. Wireless: The average rating of overall QoIS of all

ISPs was 4.01. Reliance Jio and Airtel were highest at 4.08 and 4.07 respectively. It was closely followed by Uninor and Vodafone which had ratings of 4.03 each. The lowest rating was attained by Tata at 3.59.

B. Wireline: The average rating of all ISPs on overall QoIS was 4.15. The best performer was Airtel with a mean rating of 4.35, the only one above average. Rel Comm was the lowest performer with mean ratings of 3.98.

On a conclusive note, it may be informed that in the wireline segment, local ISPs have been rated much better than the national brands overall. You Broadband, Hathway and GTPL have been rated better than national SPs like Airtel, Rel Comm, MTNL/BSNL and Tata. Spectranet, Hathway and Tikona are rated better than national brands in Metro Circle while Spidigo, Beamtele, You Broadband, Hathway, GTPL and Tikona are rated higher than national ISPs in circle A.

10.7 Analytical Comparisons

This section shows analytical comparisons of means of ratings of wireless and wireline ISPs on five parameters and overall quality of service for all circles taken together.

10.7.1 Availability: The first quality parameter is availability of internet service. It covers aspects like availability of information on plans in local languages and notification about changes in plans. Respondents were asked to rate these two aspects of availability of their internet service on a scale of 0-5, where 0 is 'can't say', 1 stands for 'poor' and 5 for 'excellent'. All respondents who answered 'can't say' on any of the individual aspects have been excluded while calculating the overall availability and so is the case for those whose rating on 'overall availability' was 'can't say' while conducting the ANOVA.

A. Wireless:

- Reliance Jio was rated highest amongst all wireless ISPs on the chart of 'overall availability' with mean values of 4.31.

- Rel Comm (4.29) and Airtel (4.27) were also rated above average (4.23). However, Idea (4.20) was slightly lower. All other ISPs were thus rated less than average.
- Uninor was noticeably low in their rating on 'overall availability' with a mean value of 3.83.

We conducted Games Howell Post-Hoc Test in continuation to identify which service providers differ significantly in their ratings on 'overall availability'. The results showed that:

- Aircel and Uninor were found to have significant difference in their mean ratings on overall availability with that of all the eight other wireless ISPs.
- Apart from Aircel and Uninor, Airtel and Reliance Jio were significantly different from each other and all other ISPs except that of Rel Comm.
- In addition to Airtel, Reliance Jio, Aircel and Uninor, Idea also differed significantly from Vodafone, Rel Comm and MTNL/BSNL.
- Rel Comm differed significantly from Vodafone, Tata and MTNL/BSNL apart from Aircel, Idea and Uninor.

B. Wireline:

- The average rating on overall availability of all wireline ISPs taken together was 4.29.
- You Broadband topped the chart of 'overall availability' with mean rating of 4.40, closely followed by Hathway (4.38) and Airtel (4.36).
- Tikona was also rated above average with a mean rating of 4.30 each. The mean rating of Alliance and GTPL was equal to average (4.29).
- Connect Broadband at 3.98 was rated lowest on this scale.

To identify which service providers were statistically significant, we conducted Games Howell. The results showed that:

- Connect was significantly different from that of 11 other wireline SPs.
- Other than Connect, Asianet differed significantly from all

other ISPs but that of Rel Comm and Tata.

- Apart from Asianet and Connect; Airtel, You Broadband and Hathway differed significantly from Rel Comm, Tata and MTNL/BSNL.

10.7.2 Speed: The second parameter of quality, but the most critical is speed of internet service. It includes aspects like downloading and uploading speed, their consistency and comparison with committed speeds. Respondents were asked to rate these aspects of speed on a scale of 0-5, where 0 is 'can't say', 1 stands for poor and 5 for excellent. All respondents who answered 'can't say' on any of the individual aspects have been excluded while calculating the overall speed and so is the case for those whose score on 'overall speed' was 'can't say' while conducting ANOVA.

A. Wireless:

- The average rating of all wireless ISPs taken together was 4.21.
- Reliance Jio obtained highest mean value of 4.29 on overall speed. It was followed by Airtel (4.26), Rel Comm (4.23) and marginally lower than average was of Idea (4.20).
- 5 other ISPs namely Vodafone (4.15), MTNL/BSNL (4.13), Tata (4.05), Aircel (4.03) and Uninor (3.85) were rated below average, of which Uninor was remarkably low.

To identify which SP is that, we conducted the Games Howell post-hoc test. This test showed that:

- The top-rated Reliance Jio and least-rated Uninor have been found to be significantly different from all the eight other SPs.
- Apart from Reliance Jio and Uninor; Airtel and Idea had significant differences in their mean ratings with each other and Aircel, Vodafone, Tata and MTNL/BSNL.
- Apart from Airtel, Idea, Reliance Jio and Uninor, Aircel significantly differed from Rel Comm, Vodafone and MTNL/BSNL and Tata only from Vodafone and Rel

Comm.

- Rel Comm differed significantly from Vodafone and MTNL/BSNL in addition to Aircel, Tata, Reliance Jio and Uninor.

The post-hoc analysis shows that there are no significant differences among Airtel and Reliance Jio—top two ISPs on speed.

B. Wireline:

- You broadband was rated highest on 'overall speed' with mean value of 4.42, closely followed by Hathway (4.38) and Alliance (4.37).
- GTPL (4.35), Airtel (4.33) and Tikona (4.31) were also rated above the average score which was 4.25.
- The least rating was obtained by Connect Broadband which was 3.97.

We conducted Games Howell post-hoc test to identify which service providers differed. The multiple comparison test indicated that:

- Connect was significantly different from all ten other ISPs except with Asianet which was significantly different from all others except Rel Comm and Tata.
- Apart from Asianet and Connect; Alliance, GTPL, Hathway, Airtel and You Broadband also differed from Rel Comm, Tata and MTNL/BSNL, but Tikona differed from Rel Comm and MTNL/BSNL.

10.7.3 Reliability: The third quality parameter is the reliability of internet service. It measures aspects like uninterrupted services at all times and places and the effectiveness of roaming services. Wireless respondents were asked to rate on all these aspects of reliability, but wireline respondents had to rate only on 'uninterrupted services at all times' as the other two are not applicable in their case. Similar scale of 0-5, where 0 is 'can't say', 1 stands for poor and 5 for excellent was adopted. All respondents who answered 'can't say' on any of the individual aspects have been excluded while calculating the overall reliability and so is the case for those whose rating on 'overall reliability' was 'can't say' while conducting the ANOVA.

A. Wireless:

- Most reliable wireless ISP was Reliance Jio with mean rating of 4.28.
- Marginally higher than the overall average rating on reliability (4.20) were Airtel and Rel Comm with mean values of 4.24 and 4.22 respectively.
- Uninor was rated as least reliable wireless ISP having an average rating of 3.87.
- The ratings of other four national level ISPs such as Idea, Vodafone, MTNL/BSNL, Tata and Aircel as reliable providers were also below the average mark.

Games Howell multiple comparison post-hoc test was used to identify ISPs which are significantly different in terms of providing reliable services. The findings of this test showed that:

- The service reliability of Uninor, rated at the bottom end, was found to be significantly different from all the eight other wireless ISPs.
- Apart from Uninor; Airtel, Idea and Reliance Jio differed significantly from each other and all five ISPs except Rel Comm.
- In addition to Airtel, Idea, Reliance Jio and Uninor, Aircel also had significant differences with MTNL/BSNL, Vodafone and Rel Comm but Tata differed from Vodafone and Rel Comm.
- Rel Comm differed from Vodafone and MTNL/BSNL in addition to Aircel, Tata and Uninor.

B. Wireline:

- National ISPs—You Broadband (4.72) and Hathway (4.58) were rated highest on 'overall reliability'. This was closely followed by GTPL that serves in Gujarat, Maharashtra and Rajasthan and Airtel, having ratings of 4.49 and 4.40 respectively.
- Government owned ISPs—MTNL/BSNL (4.24) were

rated below average of 4.31.

- Asianet and Connect were substantially lower with ratings on 'overall reliability' at 3.96 and 3.87 respectively.

To specify which ISPs act similarly, we performed the Games Howell multiple comparison test. The results gave the following impressions:

- Connect differed significantly from all nine other ISPs but Asianet differed from all others except Alliance and Tata.
- Apart from Asianet and Connect; Rel Comm and MTNL/BSNL had significant differences with Airtel, Hathway, You Broadband and GTPL, whereas Tikona differed from Hathway and You Broadband and GTPL differed from You Broadband.
- Alliance and Tata differed from Airtel, Hathway, You Broadband and GTPL in addition to Connect.
- Airtel differed from You Broadband and Hathway in addition to Rel Comm, Tata, Connect, Asianet, Alliance and MTNL/BSNL.

10.7.4 Billing and Charges: The fourth parameter of quality i.e. billing and charges of internet service is the most influential one as it determines the choice of SP. Normally, wireline service is much more costly than the wireless service due to its installation and fixed charges. It covers aspects like transparent billing and data consumption policy and charging for renewal of plans. Respondents were asked to rate these aspects of billing and charges of their internet service on a scale of 0-5, where 0 is 'can't say', 1 stands for poor and 5 for excellent. All respondents who answered 'can't say' on any of the individual aspects have been excluded while calculating the overall billing and charges and so is the case for those whose rating on 'overall billing and charges' was 'can't say' while conducting the ANOVA.

A. Wireless:

- As expected, the market-disrupter Reliance Jio topped the chart of 'overall billing and charges' with a mean rating of 4.29. However, in order to beat the competition, other wireless SPs offered new plans with much less prices. It is due to this that there are not striking differences in scores of other wireless SPs.

- The second highest rating was that of Airtel's i.e. 4.24, slightly lower than that of Reliance Jio.
- Rel Comm (4.22) was rated just above the overall average of 4.20.
- Uninor was rated lowest at 3.90 on 'overall billing and charges'.

To highlight which ones are different, we conducted the Games Howell post-hoc test, the results of which led to the following suggestions:

- The top performer Reliance Jio behaves uniquely as it has significant differences with all the eight other SPs.
- Apart from Reliance Jio, Airtel and Idea differed significantly from all other ISPs except Rel Comm, whereas Uninor and Vodafone differed with all others except Aircel and MTNL/BSNL respectively.
- In addition to Airtel, Idea, Reliance Jio and Uninor; MTNL/BSNL had significant differences with Rel Comm, which also differed from Tata and Aircel.

B. Wireline:

- You Broadband having a mean value of 4.50 has been rated highest on 'overall billing and charges'. Hathway and GTPL followed with mean ratings of 4.42 and 4.39 respectively.
- Quite close to the average rating of 4.25 were Alliance (4.33), Tikona (4.31) and Airtel (4.31).
- Connect was rated lowest for their billing and charging policies with mean values of 3.92.

Following are the highlights of the results of Games Howell post-hoc test which we have used to identify the service provider(s) that have significant difference in their charging policies:

- The lowest rated Connect was significantly different from all other ten ISPs.

- Apart from Connect; You Broadband differed significantly from all other ISPs except Hathway and GTPL; MTNL/BSNL differed from all others except Rel Comm, Tata and Asianet; and Airtel differed from all others except Tata, Tikona and Alliance.
- In addition to Connect and You Broadband, Tata differed from Hathway and GTPL whereas Rel Comm differed from Hathway, Tikona, GTPL and Alliance.
- Also, Hathway, Tikona, GTPL and Alliance differed from Asianet.

10.7.5 Help and Customer Care Services: The quality parameter—help and customer care of internet service—relates to aspects like accessibility and effectiveness of help services/customer care executives and transparency of dispute settlement mechanism and refund policy. Wireless respondents were asked to rate these aspects on a scale of 0-5, where 0 is 'can't say', 1 stands for poor and 5 for excellent. In addition to these, wireline respondents rated the provision of technicians as per the requirement. All respondents who answered 'can't say' on any of the individual aspects have been excluded while calculating the overall help/customer care services and so is the case for those whose rating on 'overall help/customer care services' was 'can't say' while conducting the ANOVA.

A. Wireless:

- Reliance Jio and Airtel were rated highest on 'overall help/customer care services' with mean values of 4.30 and 4.26 respectively.
- Rel Comm was also rated marginally above average (4.22) at 4.23.
- Uninor was rated least on the chart of overall help/customer care services at 3.93.

To identify which ISPs differ, we applied the Games Howell post-hoc test. The results of this test suggest that:

- Reliance Jio has been found to have statistically significant difference in its help service offerings with all the eight other wireless SPs, implying that it is the best among all

ISPs in this regard.

- Apart from Reliance Jio, Airtel and Idea differed from all other ISPs except Rel Comm; and Vodafone differed significantly from all others except MTNL/BSNL.
- In addition to Reliance Jio and Vodafone, Rel Comm also differed significantly from Aircel, Tata, MTNL/BSNL and Uninor. Also, Uninor differed from MTNL/BSNL.

B. Wireline:

- You Broadband was rated the highest on 'overall help/customer care services' at 4.45, followed by Hathway (4.43).
- GTPL (4.36), Alliance (4.33), Airtel (4.32) and Tikona (4.32) were rated above the average rating of 4.26.
- Connect Broadband was rated the lowest on 'overall help/customer care services' at 3.93.

Now, to identify which ISPs differ, we conducted the Games Howell post-hoc test. The test results suggest that:

- Connect differs significantly from all other ten wireline ISPs except with Asianet, which further differs from all others except Rel Comm and Tata.
- Apart from Connect and Asianet; Alliance, GTPL and Tikona differed significantly from Rel Comm, Tata and MTNL/BSNL.
- Also, in addition to Connect and Asianet, Airtel differed from Rel Comm, Tata, MTNL/BSNL, Hathway and You Broadband.
- Apart from Connect, Airtel, Alliance, GTPL and Tikona; Rel Comm and Tata differed from Hathway and You Broadband.
- Also, MTNL/BSNL differed significantly from Hathway and You Broadband.

10.7.6 Overall Quality of Service: Apart from five parameters of quality that have been discussed in detail above, a separate question that sought rating on overall quality of internet service is also included in the questionnaire.

The scale used for this rating was 1-5, where 1 stood for

poor and 5 for excellent.

A. Wireless:

- The mean value for all the wireless SPs stood at 4.14.
- Airtel and Reliance Jio with mean values of 4.25 and 4.18 respectively were the only two ISPs whose overall quality of service was above average.
- All other SPs were rated slightly below average, except Uninor that had mean rating of 3.88, which was the lowest.
- The differences in the means are significant at .05 level and so their services are not homogenous. This could be seen from the ANOVA analysis shown below.

To determine which service provider(s) were different in their quality of internet service, the means were examined through the Games-Howell post-hoc test. The application of this test led to the following results:

- Service of Airtel and Reliance Jio was significantly different than that of all the eight other wireless service providers.
- Apart from Airtel and Reliance Jio, Uninor differed significantly from all other ISPs except Aircel.
- In addition to Airtel and Reliance, Aircel had significant differences with Idea and Vodafone, which were also different from each other.

B. Wireline:

- You Broadband was ranked as the top most provider in terms of overall QoIS with an average score of 4.71. It was followed by Hathway and GTPL with mean ratings of 4.55 and 4.46 respectively.
- National ISPs Airtel (4.24) and Tikona (4.20) were also rated above average (4.17).
- All other providers were rated below average with Connect (3.95) being rated the lowest.
- The differences in the means are significant at .05 level and so their services are not homogenous.

To identify which service provider has a different service, the means were examined in order to determine which means differ from each other through the Games-Howell post-hoc test. The application of this test showed that:

- On the top end, You Broadband differed significantly from all ten other wireline SPs.
- Apart from You Broadband, Airtel differed from all other ISPs except Tikona and Asianet whereas GTPL and Hathway differed from all other ISPs except with each other.
- Apart from Airtel, Hathway, GTPL and You Broadband; Connect and Alliance differed significantly from Rel Comm, Tikona and MTNL/BSNL.

11

Speed of Internet Services

Speed of internet service is not just considered to be an important parameter for the quality of internet, rather it is also a principal factor that influences the purchase decision that a consumer takes when given a plethora of plans that are priced on the basis of combinations of speed and data allowance to choose from. In this digital era, where communication services have become an essential part of our economy, interest of consumers with respect to internet services cannot be sidelined. However, a large number of internet users in India are still unaware of the difference between the speeds they pay for and the actual speeds they get.

With the advent of new technologies, it has now become possible to provide bigger and better services to internet users. But with increasing number of internet connections, incidences of consumer detriment and dissatisfaction have also surged. TRAI acknowledges that there exists *information asymmetry* in the telecom sector which is one of the causes of market failure. It basically pertains to the situations "where one transacting party has access to more superior or accurate information compared to the other party". Non-disclosure of provisions of services like accurate data speeds, contract terms, latency, etc. may cause customer dissatisfaction. Even if the complete information is available, it might not be in a format that is easily comprehensible to the customers. Use of terms like 'up to' for data speeds has become a common occurrence although the theoretical speeds (as claimed by ISPs) are rarely delivered.

11.1 Consumer Perception about Speed

One of the recently conducted IVR based surveys by TRAI

to assess the consumer perception found that about 60 percent, 62 percent and 67 percent respondents were dissatisfied with their internet data speeds in Delhi, M.P. and Karnataka respectively. Also, satisfaction with data speeds was given a rating of 2.39, 2.33 and 2.14 on a five-point scale, in Delhi, M.P. and Karnataka respectively (where 3 was neutral, below 3 was dissatisfied and above 3 was satisfied) (TRAI Survey Reports, 2017).

After receiving a number of complaints from consumers regarding poor internet speeds, TRAI proposed to 'mandate the minimum download speeds', so as to protect the interests of consumers. But service providers opposed the proposal citing a number of constraints due to which exact speeds cannot be measured, such as limited availability of spectrum per operator in 2G/3G bands; low coverage zones such as basements, high rise buildings, tunnels, etc.; quality and type of subscriber's device, number of subscribers browsing the data services, peak/off peak time, transmission bandwidth, external interference, website behaviour etc.

In October 2016, TRAI issued directions applicable to the wireline and wireless broadband services. Wireless broadband service providers are "required to disclose the data usage limit which specifies primary technology (3G/4G) and the speed offered after that limit on its website in all advertisements". In case of wireless connections, consumers are generally not aware of the actual speeds and perceive the speed to the theoretical speeds, which are claimed by the ISPs. Theoretical speeds are the 'advertised speeds' which often greatly vary from the actual speeds. Data available at TRAI's MySpeed Portal shows that a large number of 3G subscribers throughout the country receive a download speed of less than 1mbps and may go as low as 10 kbps for some users.

Table 11.1 shows the theoretical 2G, 3G and 4G speeds (TRAI, 2017).

It can be observed that there is improvement in internet speeds with each advanced generation of cellular network.

CUTS International in their broadband labels training manual (CUTS International, 2017) discuss about the various generations of cellular networks which were designed to provide better internet services in terms of speed. These generations are explained as below.

Table 11.1: Upload and Download Speeds for Various Technologies

S. No.	Generation	Technology	Download Speed (Theoretical)	Upload Speed (Theoretical)
1.	2G	GSM/ GPRS	< 0.5 Mbps	20 kbps
2.	3G	HSDPA HSUPA HSPA+ DC-HSPA+	1.8 Mbps 7.2 Mbps 28 Mbps 42 Mbps	384 kbps 5.76 Mbps 5.76 Mbps 11 Mbps
3.	4G	LTE LTE-Advanced	100 Mbps 150 Mbps	Up to 50 mbps

Source: *Qualcomm Report: Evolution of Mobile Technologies*; (Taken from TRAI consultation paper)

1. **2G:** It came in 1991 and cell phones became digital from analog. It was primarily designed for phone calls and enabled use of text messaging.
   Then came 2.5G that enabled internet on mobile phones. Users could access GPRS. Maximum theoretical download speed under 2.5G network was 'up to 80 kbps'.
   The next upgrade was 2.75G that offered improved data transfer rates through GSM. Maximum theoretical download speed under 2.75G network was 'up to 236.8 kbps'.
2. **3G:** It was introduced in 1998 and commercially launched in Japan during 2001. It enabled faster data transmission and provided scope for video streaming, downloads etc. 3G brought the mobile broadband revolution and offered speeds 'up to 42 mbps'.

3. **4G:** It was launched in 2009 and is able to support much higher speeds and data intensive tasks as compared to 3G. The maximum theoretical 4G download speeds are given in Table 11.1.

11.2 QoS Standards for Wireless Data Services

In its efforts to ensure consumer protection, TRAI has issued "Standards of Quality of Service for Wireless Data Services Regulations, 2012" that introduced a list of QoS parameters (Table 11.2). These parameters have certain benchmarks that service providers have to comply with. Following this regulation, in 2014, TRAI introduced a clause in the regulation that mandates service providers to publish minimum download speed for the wireless data plans. Also, service providers are directed to ensure that the minimum download speed specified in their advertisements is delivered not less than 80 percent of the usage time. Table 11.2 shows the QoS parameters and benchmarks laid down in this regulation (CUTS International, 2017).

TRAI has framed its regulations in such a way that allow ISPs to mislead consumers by not assuring minimum or average speeds. ISPs may not commit to an average speed, yet certain apps such as TRAI MySpeed app, allow the users to test their network speeds. But it is difficult for the users to determine the reasons behind low network speeds. As mentioned above, TRAI proposed to make disclosure of average speeds mandatory, but it has not been able to put it into effect till date and many ISPs have opposed this. However, similar efforts have been made and were somewhat successful in other countries.

11.3 Promised Speed and Actual Speed

While conducting the situation analysis for this study, the research team observed that the most significant and recurring problem faced by consumers due to poor quality of internet was that of 'speed'. They reported battling with the inconsistent speeds throughout the day and at different locations on frequent basis. Meetings with different experts and regulatory officials made it clear that speed has been largely neglected so far in spite of it being a crucial factor in determination of quality of service.

Table 11.2: QoS Standards for Wireless Data Services in India Regulations, 2012

Name of Parameter	Benchmark
Service Activation/Provisioning	Within 4 hours with 95 percent success rate
Successful Data Transmission Download Attempts	> 80 percent
Successful Data Transmission Upload Attempts	> 75 percent
Minimum Download Speed	To be measured for each plan by the service provider and reported to TRAI
Average Throughput for Packet Data	> 75 percent of the subscribed speed
Latency	Data < 250 ms
PDP Context Activation Success Rate	≥ 95 percent
Drop Rate	≤ 5 percent
Amendment 2014	
Minimum Download Speed	≥ 80 percent of the usage time

Source: TRAI website.

These observations and discussions with stakeholders prompted the research team to conduct an audit of internet speeds on important parameters. The primary objective of this was to focus on the *speed audit* to observe the difference between '*promised speed (as claimed by service providers)*' and '*actual speed (as experienced by users)*'. The study involves the audit of wireless (mobile) internet services only. The primary reason for limiting the scope of this study to wireless mode only was because it is only in this case, that there have been huge differences between promised and actual speeds unlike in case of wireline, where striking differences between the two are rarely observed. Another reason for the same was the constraint of access to the consumer's computers or any other fixed device by the auditors.

11.4 Speed Performance and Network Strength

The application basically tests four factors that make up for 'speed performance' and 'network strength'. On opening the application, it automatically detects the SIM card or wi-fi on which internet services are currently being availed of. The user has to just click on the 'begin test' tab and it presents the results within few seconds/minutes along with date and time of test. The factors which are tested through this application are defined below (TRAI, TRAI MySpeed Portal, 2017):

1. **Upload Speed:** The upload speed is how fast you send data from your device to others. It is measured in megabits per second (mbps).
2. **Download Speed:** The download speed is how fast you can pull data from the server to your device. Most connections are designed to download much faster than upload, since the majority of online activity, like loading of web pages or streaming of videos, consists of downloads. Download speed is measured in megabits per seconds (mbps).
3. **Network Delay/Latency:** Network delay is an important design and performance characteristic of a computer network or telecommunications network. The delay of a network specifies how long it takes for a bit of data to travel across the network from one node or endpoint to another. It is typically measured in multiples or fractions of seconds (milliseconds/ms).
4. **Packet Loss:** Packet loss occurs when one or more packets of data travelling across a computer network fail to reach their destination. Packet loss is typically caused by network congestion and in most cases is the

result of poor signal/line quality. Packet loss is measured as a percentage of packets lost with respect to packets sent.

11.5 Data Analysis

In the audit, data of 2,570 readings were taken into consideration. This data was collected from four telecom circles and eight states consisting of metro circle (Delhi), circle A (Gujarat and Karnataka), circle B (Punjab, Kerala and Uttar Pradesh) and circle C (Odisha and Bihar).

The internet speeds of a total of six ISPs were measured, namely Airtel, Vodafone, Reliance Jio, Idea, MTNL and BSNL. Except MTNL and BSNL, all the SIM cards were 4G enabled. MTNL was tested only in Delhi and thus had the lowest sample size.

The tests were conducted in 3G/4G enabled smart phones. The areas where these tests were conducted were Delhi (Delhi), Allahabad (U.P.), Ahmedabad (Gujarat), Ludhiana (Punjab), Patna (Bihar), Bhubaneswar (Odisha), Bangalore (Karnataka) and Ernakulam (Kerala).

These readings were analyzed based on two dimensions: States/circle-wise and ISP-wise. These observations were put through various tests, each of which has been explained below:

1. **Mean:** It is the average of all the observations under one parameter. For example, 'average download speed'.
2. **Count:** Total number of cases/observations under one parameter.
3. **Maximum:** It is the maximum numerical value in all the observations under one parameter.
4. **Minimum:** It is the minimum numerical value in all the observations under one parameter.
5. **Percentile 25:** The 25th percentile of a list of observations is the number such that at least 25 percent of the observations are no larger than it.
6. **Percentile 75:** The 75th percentile of a list of observations is the number such that at least 75 percent of the observations are no larger than it.

11.5.1 All Circle/States-wise Analysis: 3G: The sample of 3G readings was low as most of the service providers now offer 4G services. As mentioned earlier, only BSNL and MTNL 3G SIM cards have been tested under 3G category and MTNL was tested only in Delhi. Therefore, the 3G sample size is skewed towards BSNL. The results are shown circle/state-wise.

Following observations were made:

1. **Download Speed (in mbps)**

- The average 3G download speed taking all the circles together was found to be 1.50 mbps. The highest average 3G speed was found in U.P. at 2.98 mbps, followed by 1.93 mbps in Kerala, both in circle B. Lowest were recorded in Bihar, Punjab and Delhi, where speeds were not even 1 mbps.
- Overall, the maximum 3G speed was recorded at 7.11 mbps in U.P. followed by 6.78 mbps in Karnataka. This was again the lowest in Bihar at 1.67 mbps.
- Overall, maximum 25 percent readings were above 1.84 mbps. U.P. and Kerala, both in circle B, recorded the highest speeds in this range at 4.03 mbps and 3.03 mbps respectively.

2. **Upload Speed (in mbps)**

- Overall average upload speed of 3G network was recorded at 1.26 mbps. The same was recorded highest in U.P. at 1.95 mbps, followed by 1.45 mbps in Odisha. The lowest was again recorded in Bihar and Delhi where it was less than 1 mbps.
- The overall maximum upload speed recorded was 5.87 mbps in Karnataka, followed by 5.56 mbps in Punjab. Gujarat and Kerala also closely followed with 5.43 mbps and 5.34 mbps respectively.
- Overall, maximum 25 percent readings were above 1.72 mbps.

3. **Network Delay (in ms)**

- Average network delay, taking all the states together was

recorded at 137 ms. Highest was recorded in U.P. (224 ms) followed by Gujarat (217 ms) and lowest was recorded in Karnataka at 73 ms.

- Overall, highest 'maximum network delay' recorded was 1177 ms in U.P. The least was recorded in Odisha at 216 ms.

The count of observations from 3G network which fall in the three categories of packet loss (percent) are as below.

- The overall average (mean) packet loss was recorded at 15 percent. More than 60 percent observations were equal to or less than the mean value.
- All the states except Delhi had more than 50 percent readings falling within the category of 0-15 percent packet loss.
- In case of 16-25 percent category, Odisha had the highest percentage of readings at 36.2 percent. Karnataka registered the highest number of observations in the range of 26-90 percent.

11.5.2 All Circle/States wise analysis-4G:

Following observations were made:

1. **Download Speed (in mbps)**

- The average 4G download speed obtained taking all the states together was found to be 4.10 mbps. The highest average 4G download speed was found in Kerala at 6.13 mbps followed by Odisha at 5.52 mbps. Lowest was recorded in Delhi at 2.55 mbps.
- Overall, the maximum 4G download speed was recorded at 12.70 mbps in Karnataka and 12.56 mbps in Kerala. In rest of the states, it was almost similar and closer to 10 mbps.
- Overall, at least 25 percent readings were greater than 1.76 mbps and maximum 25 percent readings were above 5.56 mbps. Both of these measures were recorded highest in Kerala at 3.45 mbps and 8.45 mbps respectively.

2. **Upload Speed (in mbps)**

- Overall average upload speed was recorded at 3.41 mbps.

The highest average 3G uploading speed was recorded in Odisha (5.46 mbps) followed by Bihar (4.52 mbps) and the lowest was recorded in Delhi and U.P., both at 2.16 mbps.

- In Odisha and Bihar, at least 25 percent readings were above 3.67 mbps and 3.46 mbps respectively, highest among all other states.

3. **Network Delay (in ms)**

- The highest 'average network delay' was recorded in Gujarat at 273 ms and Odisha at 267 ms, which was even higher than the 'overall average network delay' (178 ms). The least was recorded in Karnataka at 114 ms.
- Overall, 'maximum network delay' recorded was 1393 ms in Karnataka. The same was recorded at 457 ms, the least among all other states in both Bihar and Kerala.
- Overall, the average packet loss recorded was 14 percent. It was the highest in Odisha at 25 percent.
- In aggregate, 60.8 percent of 4G network readings, recorded packet loss in the range of 0-15 percent. In Karnataka and U.P., more than 80 percent readings fell in this category, whereas in Odisha, only 14 percent observations were less than or equal to 15 percent.
- Odisha (35.6 percent) followed by Kerala (26.3 percent), recorded the highest percentage of packet loss readings ranging from 26 percent to 90 percent. The least in this category were recorded in Punjab at only 3.3 percent.

11.5.3 ISP-wise Analysis: 3G: It has been mentioned earlier that there is huge difference in the sample size of MTNL and BSNL, owing to the fact that MTNL is only available in Delhi and Mumbai. As Mumbai has not been covered in our survey, so the analysis has been done on the data received from Delhi only.

1. **Download Speed (in mbps)**

- The overall 'mean (average) downloading speed' for 3G network was recorded at 1.50 mbps. BSNL recorded a better speed at 1.61 mbps.
- In 3G, the maximum speed recorded was 7.11 mbps

captured on a BSNL network, whereas in case of MTNL it was 5 mbps.

- Overall percentile 75 of download speed was recorded at 1.84 mbps. In case of MTNL, at least 75 percent observations were equal to or less than 0.76 mbps.

2. **Upload Speed (in mbps)**

- The overall 'mean (average) upload speed' was recorded to be 1.26 mbps. Here also, BSNL fared better than MTNL with 1.31 mbps average upload speed.
- The overall maximum upload speed was registered at 5.87 mbps on a BSNL network. For MTNL, it was 4.56 mbps.
- In case of MTNL, at least 75 percent observations were below 1 mbps.

3. **Network Delay (in ms)**

- Overall 'average network delay' was recorded to be 137 ms. For MTNL, it was 92 ms and for BSNL it was 143 ms.
- Maximum network delay reading was recorded as 1177ms on a BSNL network and 238 on a MTNL network.
- The average packet loss in case of MTNL was 20 percent and 14 percent in case of BSNL.
- 64 percent readings of BSNL network registered a packet loss of less than or equal to 15 percent. This was only 36.6 percent in case of MTNL.
- One-third observations on MTNL network recorded packet loss in the range of 16-25 percent, whereas BSNL recorded only 20 percent in the same range.

11.5.4 ISP-wise Analysis: 4G:

1. **Download Speed (in mbps)**

- The overall 'mean (average) downloading speed' for 4G was recorded at 4.10 mbps. Only Airtel recorded a better speed than this at 6.05 mbps. Least was recorded on an Idea network at 2.85 mbps.
- The 'overall maximum download speed' was recorded at 12.70 mbps on Airtel network. It was closely followed by Reliance Jio and Idea at 12.23 mbps and 12.05 mbps, respectively.

- Overall 'percentile 75' of download speed was recorded at 5.56 mbps and highest was again Airtel (9.45 mbps).

2. **Upload Speed (in mbps)**

- The overall 'mean (average) upload speed' was recorded to be 3.41 mbps. Only Airtel and Reliance Jio performed better than the overall at 4.41 mbps and 3.64 mbps respectively.
- The overall maximum upload speed was registered at 9.85 mbps on Airtel network. Others followed closely.

3. **Network Delay (in ms)**

- Overall 'average network delay' in 4G was recorded to be 178 ms. The highest was recorded on Airtel network (201 ms) which is a little higher than the average.
- Maximum network delay reading was recorded as 1393ms in Idea and the least was recorded in Airtel at 1134 ms.
- The overall average packet loss recorded was 14 percent. Airtel recorded a little higher than average at 17 percent.
- More than 60 percent of total 4G observations recorded a packet loss of up to 15 percent. Vodafone had the highest proportion of observations falling into this category at 65 percent.
- 3 out of every 10 observations in Airtel network recorded a packet loss in the range of 26-90 percent, higher than any other ISP.

Looking at the above observations, we may conclude that BSNL was a much better performer in 3G category as compared to MTNL. In case of 4G network, Airtel was the best performer in download and upload speed, but also registered higher average network delays and packet loss. It was followed by Reliance Jio in download speed and by Vodafone in upload speed.

11.6 Findings and Conclusion

The loose definition specifying the range of 3G and 4G speeds in India as defined by TRAI has already been mentioned in the beginning of this chapter. However, some recent reports give an idea about the actual 3G and 4G speeds being experienced by users in India. Average 3G download speeds across the country were found to be less than 1 mbps and average 4G download speeds were found to be 5.1 mbps (Times, 2017). This corroborates the findings of our study that none of the providers are actually giving realistic 4G speeds.

The inflated speeds are advertised to lure customers to buy new and costly plans, even if they do not get the desired benefits. This was also observed when consumers were asked during the survey, about the most frequent problems they face with respect to their internet survey. 'Speed related problems' and 'poor connectivity' were the most recurring problems faced by consumers *(see chapter 12 on Consumer Problems, Complaints and Redressal Mechanism).* While buying a new SIM card offline or even recharging through ISP's website, customer is never made aware of the actual speeds he/she shall be getting. These plans are sold just mentioning the 3G and 4G speeds, no specific range of speed is mentioned. A person can rely on only the loose definition of 3G and 4G defined by the regulator. Our results show that both 3G and 4G networks fall short of expectations when tested under normal conditions. By normal conditions we mean, the state, place and time at which a consumer tries to access the internet. However, the service providers provide alibis for poor speed and connectivity. They claim that actual speeds cannot be measured due to the following constraints (TRAI, Data Speed under Wireless Broadband Plans, 2017).

1. Mobile broadband speed depends on various factors (technical mostly).
2. Mobile broadband is different from fixed broadband so a user cannot expect the same consistency in speed.
3. Location of the customer and distance from the base transceiver station (BTS).
4. Times of access—peak time/off peak time.
5. Kind of device being used.
6. Other technical aspects.

TRAI has left the regulations loosely defined and these regulations have not seriously addressed the loopholes that the ISPs use as their alibi for poor quality. All these concerns were also raised during the TRAI OHD on one of its consultation papers, held on October 3, 2017 in New Delhi. For a general consumer and that too in India where we are already crippled with low levels of awareness and education, giving these arguments as the cause of low speed, may not make much difference in the perception and expectation of a consumer. When a person is accessing the internet, all he/she is concerned about is network reliability and speed. What goes behind is none of his concern. Then why not advertise the actual speeds, is the question worth asking.

11.7 Recommendations

1. TRAI must mandate every ISP (internet service provider) to disclose a minimum average download speed that must be provided at least 90 percent of the time. For this, the regulator should direct each service provider:

- to include a speed measurement tool in their respective apps which should maintain a record of average speed during the day, week and month.
- to disclose minimum and maximum speeds on daily, weekly, monthly and quarterly basis.
- to compensate consumer for failing to adhere to average download speeds and down time.

2. There should be a consumer complaint feature in each ISPs app, where the subscribers can easily lodge their complaints and send periodic reminders till the time the resolution takes place.
3. TRAI MySpeed app, which was used for this study, should be made more robust and user-friendly. It should be able to provide speed comparisons for each BTS or the location where the consumer is using the internet. The user should be able to get daily, weekly, monthly and quarterly comparisons of speed at specific BTS which can provide

information on performance of each ISP at the same area/location.

12

Consumer Problems, Complaints and Redressal Mechanism

Right from the time of purchase of a connection, commencing use of an internet service till the time of completion of contract of service, the user of a service may face some problems. Redressal or solution for these problems can be sought through complaint registration. The process of resolving these problems, called the grievance redressal mechanism, which is majorly done through help and customer care centre of a service provider plays an important role in the assessment of quality of service.

TRAI has mandated a grievance redressal system through its regulations that each ISP must set up such a system which has a customer care service through call centres and appellate authority to resolve grievances related to the deficiency in service. Question number 34 of the questionnaire was framed and utilized in order to capture the experiences of the respondents in this context. The design of the question was such that every respondent who had a problem and had marked 'yes' against it, could go ahead and mark 'yes' or 'no' in the next two columns i.e. 'complained about it?' and 'received a response?'.

However, the status of problem resolution had to be told irrespective of the fact that it was complained about or not. A respondent who had not faced any problem with their service could simply mark 'no problem' and moved ahead to the next question. On the whole, six types of problems were identified, wherein the first one was 'activation of service' in case of wireless respondents and 'installation problem' in case of wireline respondents. This was followed by 'poor connectivity', 'disconnection of services without information', 'speed related problem', 'false promises', 'over charging' and 'any other problems'.

12.1 Summary of Problems Experienced with Internet Service

12.1.1 Telecom Circle-wise:

- Out of the total sample covered during the field work, 35,503 respondents reported wireless as their main mode of using internet. About one-third (30.9 percent) of them recorded that they have had faced at least one problem with their internet service providers (ISPs).
- Circle-wise analysis shows that more than half i.e. 54.4 percent of respondents reported at least one problem in circle C, which is also the highest percentage of such respondents. This was followed by metro (34.3 percent) and circle B (31.7 percent). In circle A, this was reported by almost one out of four (23.7 percent) respondents.
- Altogether 3,174 (28.9 percent) respondents of the 10,977 who had reported problems in their internet service had registered complaints for these problems. It is notable to see that more than two-third did not bother to register their complaints. In metro circle, every third respondent, who faced a problem, reported that he/she had put a complaint with their service providers. In contrast, in circle C, only one-fifth (21.2 percent) of the respondents, who had faced any problem in internet service, had put up their complaints.
- Almost three of four (72.5 percent) respondents who sought redressal for the problem in their service through a complaint got a response. This proportion was more or less same in every circle but at 83.2 percent in metro circle to 61.2 percent in circle A. This shows that the ISPs have installed a grievance handling system.
- Of all the respondents who had recalled to have faced at least one problem, around 65 percent reported that their problems got completely solved, whereas 20 percent had

partial resolution and other 15 percent reported that their problems were not solved at all. This shows that four out of five complainants got their problems fully or partially resolved.

- Complete resolution was reported highest in circle A (68.7 percent) but it was the least in circle C (44.3 percent).
- Circle B reported the highest percentage of respondents whose problem was not solved at all at 17.5 percent closely followed by circle C (17.3 percent).

B. Wireline:

- In the case of 16,497 wireline subscribers covered during the field work, every three out of ten (29.1 percent) respondents reported that they have had some kind of problem with their internet service provider.
- Circle-wise comparison shows that exactly two-fifth (40.0 percent) of the respondents in circle C reported at least one problem they had with their internet service provider, followed by those in metro circle (34.6 percent). In circle A and circle B, this was reported by 28 percent and 26.5 percent of the respondents respectively.
- Out of the 4,806 respondents who reported to have experienced some problem, 2,200 (45.8 percent) had complained for their problems. This was reported highest in metro circle (61.0 percent), followed by circle B (51.0 percent), circle C (41.4 percent) and circle A (37.3 percent).
- Largely, three out of four (74.5 percent) respondents who had complained about their problems had got a response to it. This trend was found to be more or less similar almost in all circles, except in metro circle where it was recorded highest at 87.0 percent.
- Of the 4,806 respondents who had reported that they had faced problem with the service, more than two-third (67.8 percent) recalled that their problems were completely solved—recorded highest in circle A as every three of out four (75.7 percent) confirmed that their problems were

completely resolved. Circle A was closely followed by metro circle (71.0 percent).

- 14.7 percent wireline customers reported that his problem was not solved at all. This was also reported highest in circle B (19.3 percent).

12.1.2 ISP-wise:

- Altogether 30.9 percent (10,977) of the 35,503 respondents, who reported wireless as their main mode of using internet, have had a problem in their internet service. The largest proportion of such respondents happened to be the users of Uninor (50.5 percent), closely followed by Aircel (48.8 percent) and Rel Comm (39.4 percent). Noticeably, the lowest percentage of problem-facing respondents (23.7 percent) happened to be users of the newest ISP i.e. Reliance Jio, which indicates that shorter the user experience, lower will be the deficiency in service or alternatively the service is efficient.
- Almost every one in three (28.9 percent) respondents having a problem had complained to their ISPs about it. Mysteriously, in case of Uninor that had the largest proportion of users having a problem, only 16.7 percent filed a complaint.
- Almost three in four complaints (72.5 percent) were responded by the ISPs. However, only 63.9 percent users of Uninor claimed that they got a response, whereas 79.8 percent users of Aircel reported so.
- All respondents who had experienced problems detailed that their problems were resolved completely, partially or not at all in the ratio of 13:4:3 approximately. But, for Uninor and Aircel, the percentage of users whose problems were completely solved was much lesser at being 31 percent and 52.2 percent respectively. Also, in case of Uninor, about half of the users (43.2 percent) that experienced problem in their service opined there problems were not solved at all.

A. Wireline:

- Almost one-third (29.1 percent) of the 16,497 respondents, who claimed that wireline is their main mode of using internet, reported that they have had at least one problem in their internet connection. The highest such fractions of users were of Tata (49.3 percent), Alliance—serving in West Bengal (46.8 percent) and Rel Comm (42.7 percent). But, only 9.1 percent of respondents who used Asianet (serving in Kerala) remarked so. This is a reflection on the relative quality of service of these brands. It is unfortunate that TRAI does not mandate each ISP to publish their data on consumer complaints and all metrics related to these aspects so that consumers can understand how their complaints are handled by a given ISP.
- About 45.8 percent of the 4,806 respondents who had faced problems, lodged a complaint and sought a solution. However, it is surprising to know that the lowest such percentage of respondents were users of Tata (36.3 percent), Alliance (34.6 percent) and Rel Comm (33.4 percent) itself. As seen above, only 9.1 percent of Asianet users reported facing a problem, yet they were the highest proportion of complainants (82.4 percent). This suggests that net users in Kerala were more active in protecting their interests as much larger share of those having problems made complaints. This may be due to civic and political awareness in Kerala.
- Of the 2,200 respondents who had complained, three-fourth (74.5 percent) received a response. In the case of Asianet, all complainants (100 percent) confirmed to receiving a response from the service provider. This was followed by the users of Connect (97.1 percent) that serves in Haryana, Punjab and Himachal Pradesh and You Broadband (92.9 percent). In contrast, less than one-fourth (23.1 percent) of GTPL (serving in Rajasthan, Gujarat and Maharashtra) complainants, confirmed receiving any response. This indicates that the customer service of

Asianet, Connect and You Broadband are much better than that of other ISPs.

- More than two-third of the respondents (67.8 percent) who reported to have problems with their ISP told that their problems were completely solved. But, such users of Hathway (56.8 percent) and Asianet (50.0 percent) were least in proportion.
- In the entirety, 14.7 percent who have had problems, exclaimed that their issues were left unresolved. Noticeably, none of the users of regional provider Alliance who have had problems reported that their problems were left completely unresolved. Whereas almost one-third users having problems in their internet service provided by national players like Hathway (26.9 percent) and MTNL/BSNL (18.3 percent) asserted that their problems were not solved at all. This suggests that regional providers appear to be more sensitive to the complaints by their customers.

12.2 Frequently Occurring Problems (FoPs)

12.2.1 Telecom Circle-wise: Amongst the 10,977 wireless respondents who reported of having at least one problem with their ISP, 'speed related problem' was found to be most frequently occurring problem (FoP), as it was reported by 5,360 (48.8 percent) wireless subscribers covered during the field work. The second most FoPs was 'poor connectivity' reported by another half of the respondents (47.1 percent), followed by 'activation of service (40.0 percent) and 'disconnection of service without information' (29.1 percent). The other FOPs— 'over-charging' and 'false promises' were also reported by around two-fifth of those who had recorded some problem in their wireless connections. This shows that almost half of the problems were related to poor speed of internet and the poor connectivity. This is an adequate evidence to conclude that quality of ISPs leave a lot to be desired.

When we look into each circle individually for these FoPs,

the trends have been sharply different except for metro and circle A that seemed to follow the trend, and it can be said that:

- The most frequent and second most frequent problems overall, switched places in circle B. Every one of two respondents reported to have 'poor connectivity' (52.1 percent) and every two of the five respondents faced 'speed related issues' (40.2 percent).
- The story of Circle C is strikingly different as it seems that in there getting a connection and its uninterrupted use is a problem in itself. These can be deduced as incidences of problematic 'activation of service' and 'disconnection of services without information' which were reported by 55 percent and 42.4 percent respondents respectively, making them the most common problems. The percentage of respondents facing 'speed related problems' (41.4 percent) and 'poor connectivity' (28.8 percent) were also quite noticeable.

B. Wireline: In case of 4,806 wireline respondents that remarked having a problem with their internet service provider, the most frequent problem was 'speed related' as it was reported 2,278 times i.e. by almost half of the respondents (47.4 percent). This was followed by 'poor connectivity' faced by every two out of the five respondents (41.4 percent) and 'installation problem' faced by more than one-third of the respondents (37.0 percent). 'Disconnection of services without information' (24.3 percent) was also reported by almost a quarter of the respondents each who had faced the problems. The least faced problems were 'over-charging' and 'false promises' as they were faced by 21.2 percent and 17.6 percent of the respondents respectively who reported to have this problem.

Circle-wise analysis of the FoPs for wireline users indicates that except for circle C which followed a completely different trend as that of all circles taken together, the others were quite similar as:

- For all four telecom circles individually as well, the least faced problem was that of 'false promises' as reported by

17.6 percent respondents overall. This is other than the problems of speed and poor connectivity.

- Circle C showed quite contrasting results from what has been seen in other circles. As seen in the case of wireless subscribers, the most frequent problem was 'installation problem' being faced by 43.7 percent of the respondents. Also, every six in ten respondents who have had problems reported either of the two—'poor connectivity' (34.2 percent) or 'disconnection of services without information' (30.3 percent).

12.2.2 ISP-wise:

A. Wireless: Out of the 10,977 wireless respondents that claimed of having at least one problem with their ISP, the most frequent problem was 'speed related problems' as it was reported by 48.8 percent of the respondents. Reported by almost half of such respondents (47.1 percent), 'poor connectivity' was identified to be the second most frequently occurring problem. It was followed by 'activation of service' that was reportedly faced by 40 percent of the wireless subscribers that had experienced problems. All other FoPs were reported by quarter or more percentage of respondents who have faced problems. Getting cheated by 'false promises' was the least occurring problem as it was marked by only one out of five (19.8 percent) respondents who had a problem.

The ISP-wise comparisons of these problems suggest that:

- The top three frequent problems were same as reported by users of all providers except Uninor. For this ISP, the percentage of respondents reporting to have 'disconnection of services without information' (60.9 percent) was greater than those facing 'activation of service' (52.6 percent) and 'poor connectivity' (51.6 percent).
- Users of Uninor who have experienced problems were in highest proportion amongst those who reported all six types of problems except 'poor connectivity', 'false promises' and 'over-charging'; for the former one—Rel Comm users who had encountered problems were highest

at 59.4 percent and for the latter two—MTNL/BSNL problem experiencing users were highest at 30.0 percent and 27.8 percent respectively.

B. Wireline: From the 4,806 wireline respondents that reportedly faced at least one problem in their internet service, the most common problems were 'speed related' and 'poor connectivity' as 47.4 percent and 41.4 percent such respondents reported them respectively. On the other hand, the least common problems were 'cver-charging' (21.2 percent) and 'false promises' (17.6 percent) with marginal differences.

ISP-wise analysis of frequency of problems shows that the users who faced problems of Airtel, MTNL/BSNL, Hathway and Connect only followed the somewhat similar pattern as that of all wireline respondents taken together and the rest were different as it can be seen that:

- In case of Rel Comm, Tata and Tikona, the most common problem was 'installation problem' as it was reported by 51.7 percent, 48.7 percent and 50 percent of the problem experiencing users respectively. It was followed by 'speed related issues' and 'poor connectivity'.
- For You Broadband and Alliance users who had experienced at least one problem, the most frequently occurring problem was 'poor connectivity' as reported by 59.3 percent and 46.9 percent respectively.
- The largest fractions of users having faced each problem individually were that of GTPL for all six problems except 'installation problem' where GTPL users were least in proportion. This indicates that it was quite easy to get a GTPL connection installed, however the user shall be troubled at every stage thereafter.
- The users of Asianet did not experience 'false promises' and 'over-charging' at all. This indicates exceptional conduct by Asianet which is uncommon in this industry.

12.3 Conclusion

On the whole, 31 out of 100 subscribers had a problem

with wireless internet. However, this number goes up to 1 in 2 in circle C. Considering the fact that there are hundreds of millions of subscribers; we are talking about problems faced by millions of subscribers who have been left to fend for themselves by the government and the TRAI. Those subscribers are literally at the mercy of the ISPs. The problem is corresponded by the fact that seven out of ten of such subscribers do not log any complaint. This is the indicative ease of lodging complaints with the ISPs. It is not at all easy to log any complaint and follow them up. It is true that six out of eight who lodged a complaint did not get a response. Moreover, 12.5 percent to 17.5 percent of those who complained did not get their problem resolved, indicating service deficiency.

This is a good evidence to prove that the complaint handling systems prescribed by the TRAI for ISPs which requires a benchmark of 95 percent complaints to be solved is clearly being violated. The responsibility for this situation rests on the shoulder of TRAI which is not monitoring the ISP complaint handling systems and their regulations. Since 2015, the TRAI has also stopped carrying out independent consumer surveys which can reveal such glaring violation of regulations set by the TRAI.

13

Consumer Detriment: Time Loss and Money Loss

Consumer detriment (CD) refers to the loss or harm suffered by a consumer when her/his experience with a product or service falls short of expectation. The loss can be monetary or non-monetary and can be caused due to market malpractices, defective products, deficient services, among other problems that consumer might face. In the context of the QoIS study, consumers would face losses when the quality of internet service delivered is sub-standard. Although poor quality of internet can cause consumer detriment in many ways, this study focuses on only two aspects: time loss and money loss. This was also done for easy understanding of respondents, who might be able to easily associate losses in terms of time and money on account of deficient QoIS.

13.1 Consumer Detriment: Telecom Circle-wise

- Out of the total sample of 52,000 respondents, 12,048 (23.2 percent) respondents, approximately one-fourth of the total sample, reported experiencing some or the other type of consumer detriment (CD) in recent past. Therefore, altogether 39,952 (76.8 percent) respondents had not experienced any loss due to poor quality of internet service.
- Overall, urban as well as rural respondents registered almost same percentages of the incidences of consumer detriment.
- When the responses were examined according to the telecom circles, the highest percentage of people (both rural and urban) who experienced CD belonged to circle

C, where three out of ten (30.9 percent) respondents interviewed reported to have experienced consumer detriment (CD). The lowest percentage was observed in circle A, where only 19.4 percent reported to have faced any kind of loss. In metro circle and circle B, one-fourth (25 percent) of the sample reported the instances of consumer detriment (CD).

- Out of the total respondents (12,048) who had faced CD, nearly three out of five (58.7 percent) reported 'loss of time', about one-fifth reported 'loss of money' (19.5 percent) and 'time and money loss' (21.8 percent).
- Area-wise response shows that a comparatively higher percentage of 'rural people' faced 'loss of time' (65.5 percent) as compared to urban respondents (57.1 percent). Circle C was an exception where urban respondents (69.7 percent) reporting detriment were found to be slightly higher than their rural counterparts (68 percent).
- In the case of 'loss of money', 21 percent urban respondents suffered monetary losses as compared to 13 percent of rural respondents. However, there was not much difference observed in urban and rural population in the case of 'loss of both time and money'.
- Amongst all the telecom circles, highest percentage of respondents having experienced 'loss of time' was found in circle C (69.2 percent) and highest proportion of those who experienced 'loss of money' was found in circle A (31.6 percent). Metro circle reported the highest proportion of those who faced both the losses (time and money) (23.6 percent).
- The highest percentage of rural respondents facing 'loss of time' was reported from metro circle (75.9 percent) and highest percentage of urban respondents facing the same issue belonged to circle C (69.7 percent).

13.2 Consumer Detriment: Age and Gender-wise

- Gender-wise analysis shows that the incidence of

consumer detriment was marginally higher among women than men, as 24.7 percent of the women reported to have experienced consumer detriment (CD) as against 22.9 percent of men.

- Although women have reported greater instances (65.2 percent) of 'loss of time' than men (57.6 percent); men reported to have suffered 'loss of money' and loss of 'both time and money', more often than women.
- Age-wise analysis shows that the incidence of consumer detriment (CD) was reported higher by middle aged people (36-45 years) due to poor quality of internet service. This may be due to the fact that people in this age group have more responsibilities and any loss of time and money pinches them the most.
- When looked specifically at 'time loss', highest percentage was reported by people aged between 14-35 years (61 percent), followed by respondents from the age group of 36-45 years (55.4 percent). These age groups comprise of students and the matured workforce of our country, who predominantly depend on the internet for their studies and also have importance of quality of internet service and could apprehend the 'loss of time' in their work on account of poor quality of internet service.

13.3 Consumer Detriment: Education and Profession-wise

- When analyzed according to the education qualification, the highest percentage of respondents experiencing consumer detriment due to poor quality of internet was found to be graduates and post-graduate professionals.
- The lowest incidence of consumer detriment (CD) due to poor quality of internet was reported by HSC/SSC pass respondents, yet they are the ones reporting second highest percentage of 'time loss' (64.7 percent), which is closely preceded by graduation/diploma holders pursuing the same (65.1 percent).
- The highest percentage of incidents involving 'money

loss' was reported by respondents who were graduates/post-graduates (about 22.3 percent), followed by graduate/post-graduate professionals (18.6 percent).

- Profession-wise, more than one-fourth of respondents working in the 'government/public sector' experienced consumer detriment, highest in the working class category. Respondents falling in 'others' (55.5 percent) category, which mostly includes farmers, sales persons, managers, clerks, etc. reported highest percentage of experiencing consumer detriment.
- Housewives reported the highest percentage of 'time loss' (67.9 percent) and government employees (24.7 percent) reported the highest percentage for 'money loss', after 'others' (34.7 percent).
- As for 'time and money loss', highest percentage was reported by retired and unemployed people (27.5 percent), followed by respondents engaged in private service (22.9 percent).

13.4 Consumer Detriment: SEC-wise

- Looking at the SEC analysis, it can be concluded that consumer detriment is experienced more by the lower socio-economic classes as compared to the affluent and middle class of the society. The incidence of loss rises along the SEC continuum from SEC A to SEC E, except for the middle class (SEC C), where it is the lowest.
- 28 percent individuals from lower class (SEC D and E) suffered some loss due to poor quality of internet as compared to only 19.8 percent respondents from SEC C.
- However, it can be noted that the type of loss faced by respondents from among various socio-economic classes had a variation. SEC B and D reported higher 'loss of time', whereas the affluent class reported highest percentage of 'money loss' instances (20.7 percent).
- Highest percentage of incidences of 'time and money loss' was reported by lower class respondents (24.5 percent)

and least was reported by the affluent class (21.4 percent).

13.5 Conclusion

Consumers face detriment due to poor quality of internet service in terms of 'time loss' or 'money loss' or both of these. The largest share is of consumers facing 'time loss', while the 'money loss' and 'time and money loss' has been experienced by 20 percent consumers in each category. In total, about one-fourth (23.2 percent) of the respondents had experienced at least one type of consumer detriment. This is just out of a sample of 52,000 respondents; the number could be in millions considering all the internet users in the country. It is for TRAI to design methods to eliminate such consumer detriment and draw up means of compensating consumers where they face loss of time or money on account of any business conduct by the ISPs.

14

Recommendations Based on the Findings of VOICE Survey

Based on the findings of this study, VOICE makes the following recommendations to the TRAI and the Government of India.

14.1 Resume Assessment of QoS through Third Party Assessment

The framework of Telecom Regulatory Authority of India (TRAI), established in 1997 is to "lay down the standards of quality of service to be provided by the operator and ensure the quality of service and conduct a periodical survey of such services provided by the operators so as to protect the interest of the customers of telecommunication service". In this concern, TRAI initiated study to assess the quality of telecom services every year, by third party agencies, selected through tender notifications. The study, through customer satisfaction survey and audit of the operator's operation, covered the whole of India comprising of four metros and three circles, initially for basic wireline and cellular services only and later included the broadband service. The result of these annual studies used to be regularly published through press statements and also the full report used to be uploaded on the website. This used to assure minimum level of customer satisfaction with quality of services provided by operators of these three services.

However, in April 2015, after inviting the tenders from various agencies, the process of assessing customer perception on quality of service through customer satisfaction survey (CSS) was discontinued without providing any reason. Considering the fact that individual customers cannot seek redressal from TRAI, CSS was a powerful tool for understanding the quality of operators' services on different parameters. This provided an option for the telecom consumers to have the right choice. Therefore, VOICE would recommend TRAI to reconvene the practice of assessing quality of telecom service (QoTS) through annual CSS.

14.2 Compensate for Loss of Time and Money

Since poor quality of internet service causes loss of time and money to consumers, telecos/ISPs should be made responsible for causing consumer detriment and provide monetary compensation for detriment caused to consumers. TRAI/GoI should create norms on the basis of which consumers should be compensated monetarily. In case there are legal impediments, the government must amend the relevant legislation to specifically empower TRAI to set up legally enforceable mechanism to ensure compensation of consumers.

14.3 Simpler and Easier Consumer Complaint System

A large number of consumers face problem with ISPs. However, all of them do not make complaints. This is mainly because they either feel that there is no point in wasting time with complaints or it is not easy to log complaints. Hence, lodging of complaints must be made simple and easy. Most ISPs have launched their apps which are being used to promote their sales and collection of payments from consumers. However, they do not contain a feature to log complaints. TRAI should mandate that each app must have a complaint feature with follow up of outcome of complaints. This feature should be integrated with ISPs mechanism of complaint handling. This feature should be included in each ISPs app, where the subscribers can easily lodge their complaints and send periodic reminders till the time the resolution takes place.

14.4 Promote Internet Applications among Rural and Lower Income Consumers

Internet appears to be a bridge to cross the urban-rural

divide. However, this report finds that rural consumers are way behind their urban counterparts in appreciating the potential of internet to change their lives. Hence, government policy must ensure good quality for speed. Moreover, in rural areas, digital user education programs tailored to enhance appreciation of internet tools for rural consumers need to be launched by the appropriate institutions.

Similarly, the lower income users in urban areas are way behind affluent and middle class consumers. Therefore, user digital education programs should be promoted for lower income citizens by the appropriate institutions.

14.5 E-governance

This study has demonstrated evidence that poor quality internet service can disrupt delivery of e-governance programmes. There is need to ensure that reliable fast speed internet is available for citizens to benefit from e-governance services. Whenever, any e-governance service requires filing of forms or detail of any sort, the software should include *autosave* features to save data while availing this service, to avoid loss of data due to reduced speed or disconnection.

14.6 Mandatory Minimum Download Speed

TRAI must mandate every internet service provider (ISP) to disclose a minimum average download speed that must be provided at least 95 percent of the time. For this, the regulator should direct each service provider as under:

- To include a speed measurement tool in their respective apps which should maintain a record of average speed during the day, week and month.
- To disclose minimum and maximum speeds on daily, weekly, monthly and quarterly basis.
- To compensate consumers for failing to adhere to average download speeds and down time.
- To declare the minimum and average speed at the time of sale of their plan. It must become obligatory to declare

their minimum speeds at the time of communication of their plan to consumers.

14.7 Robust and User-friendly TRAI MySpeed app

TRAI MySpeed app, which was used for this study, should be made more robust and user-friendly. It should be able to provide speed comparisons for each base transceiver station (BTS) or the location where the consumer is using the internet. The user should be able to get daily, weekly, monthly and quarterly comparisons of speed at specific BTS which can provide information on performance of each ISP at the same area/location.

14.8 Information about Signal Loss

TRAI should find a way so that consumer using internet can come to know the reduction in speed or loss of signal at a particular point of time. At present, the phones show 'E' or '3G' or '4G' signal when there is in fact no signal. When signal is lost there should be a way to inform the consumer about loss of speed or signal by an adequate label mechanism.

15

Challenges, Risks and Cyber Security

Information technology (IT) is one of the critical sectors that rides on and resides in cyber space. It has emerged as one of the most significant growth catalysts for the Indian economy. In addition to fuelling India's economy, this sector is also positively influencing the lives of the people through direct and indirect contribution to the various socio-economic parameters such as employment, standard of living and diversity among others. The sector has played a significant role in transforming India's image to that of a global player in providing world-class technology solutions and IT business services. The government has been a key driver for increased adoption of IT-based products and IT-enabled services.

However, internet poses a slew of challenges and risks to various stakeholders. Internet quality parameters, as identified by various studies, can be gauged only after a threshold of uninterrupted internet access is ensured. In other words, internet access is a prerequisite to quality of internet services. However, it must be noted that internet access is not something only internet service providers can be held responsible for. The ISPs cannot alone ensure uninterrupted internet supply when the government acts otherwise.

15.1 Challenges of Expanding Internet Base

Telecommunication companies encounter the challenge of increasing user base in a particular area with a limited bandwidth which leads to lower speed resulting in shrunken user base as telecommunication companies find it difficult to maintain internet quality as the user base expands. This is one of the reasons why average internet speed in India is lower than all other BRICS economies and lower than most other emerging economies. On the occasion of World Wi-Fi Day (June 20, 2016), an article was published in Economic Times, which discussed the plight of internet users in India. Wired broadband speeds available to customers in India is way lower than what is promised by the service providers. A data service pack that guarantees a speed of 8 mbps typically provides 5 mbps at maximum. Wireless connections are much worse. TRAI has a conservative definition of broadband that download speed should not fall below 512 kbps. That gives customers quite a narrow window to complain (Parbat, 2016).

Elucidating on the TRAI's mandate, an article titled "TRAI's meagre 512 kbps broadband mandate will push India into a technological slump" was published by the Huffington Post. TRAI issued a mandate stating that all fixed line internet service providers will provide a minimum speed of 512 kbps to their users in order to be classified as a broadband service provider. The mandate specifies that instead of the earlier 64 kbps download speed, users will now get a minimum download speed of 512 kbps after they have used up their data at higher speeds. The speed of 512 kbps for a broadband connection however, is disappointing, especially when the government is trying to boost its start-up ecosystem and promoting initiatives such as Digital India.

Another challenge encountered by the TRAI is with regard to VoIP. VoIP enables customers to make calls to any phone number in the country, be it a landline or mobile number, the party being called need not even have any app, or even internet connection for that matter. This would enable customers to make calls over the internet at one-third the current calling rates and has numerous advantages such as the ability to make calls over wi-fi in bad signal areas, 90 percent cheaper international calling, HD voice, and innovations like simultaneous data transmission, video calls etc.

In telecom license, the Government of India in 2008 had made provisions permitting unrestricted internet telephony which enables any telecom carrier in the country to provide lower cost calls through internet to any mobile or landline number in the country. The incumbent players, in response to TRAI's consultation paper on VoIP countered the "Free Your Voice" campaign, a national campaign launched by entrepreneur Bhavin Turakhia, aimed at freeing the voice of the people of India. The incumbent telecommunication companies claimed that only the telecom carrier that provides a data connection to a subscriber should be allowed to offer internet telephony to that subscriber over their internet connection.

Bringing in net neutrality is another such critical issue. From the consumer perspective, net neutrality can be defined as the principle that all electronic communication passing through a network is treated equally, independent of content, application, service, device, source or target. Consumers expect the ISPs to provide them access to the pool of resources and applications. They also expect them to comply with the fundamental principles of openness, inter-operability and neutrality. However, some operators providing end-user connections undermine users' rights when they undertake certain discriminatory activities, thereby challenging the neutral architecture of the internet. A solution suggested by some to this problem is switching to some other ISP. However, it is not a viable solution always. Switching to a better service provider often comes at a cost. Switching costs include contract cancellation fees, costs related to setting up the new network and installation costs, as well as those related to informing third parties of the new contract information. Moreover, switching may not be possible for customers who are confronted with a limited choice of operators available in a specific geographic area.

15.2 Risks Posed by Internet

Cyber security is a matter of great concern. Albeit, the Government of India and the TRAI keep coming up with ways to combat cyber crimes, the number of cyber crimes is growing. The Indian Government is reinforcing systems and establishing a centre for cyber security.

Ravi Shankar Prasad, in an interview, stated that India has about 440 crore internet users. Out of them, every 1 in 2 connections is mobile-based. The number of mobile phones added in India in the past two years, is equal to the combined population of France and Italy. Government has launched schemes such as e-health, e-scholarship for students, e-mandi for farmers, digital lockers, Jeevan Pramaan for pensions, MyGov for citizen engagement etc. Government has taken various steps to minimize cyber security threats in India including training police personnel, judges, and appointing security auditors. Also, the government has urged big companies such as Facebook, Microsoft, and Google to set up their data centres in India.

Cyber terrorism is yet another threat that is severely affecting the quality of internet services. According to a news report (Press Trust of India, 2016), ISPs in Maharashtra claimed to have suffered DDoS (distributed denial of service) attacks. A denial of service attack typically bombards websites with requests, overloading the portal until its server crashes, thus denying access of the website to legitimate users. Such attacks reduce internet speed to almost zero.

The United Nations enacted a resolution which supports the "promotion, protection and enjoyment of human rights on the internet". The resolution specifically "condemns unequivocal measures to intentionally prevent or disrupt access to or dissemination of information online in violation of international human rights law and calls on all States to refrain from and cease such measure".

According to a report by the Centre for Communication Governance at National Law University, Delhi, "internet shutdown issues relate to access to the internet and internet services at all times. The Indian government has routinely cut access to the internet during times of public strife; the rationale being that internet services (like texting apps, social media etc.) are used to coordinate and spread violence".

There have been many instances of shut downs in a number of countries across the world wherein countries block particular applications, turn off mobile telecommunications services, shut down specific services like instant messaging, and voice over internet protocol (VoIP) calling, or disrupt the entire internet. These actions separate people from their friends and family, affect their livelihoods, undermine economic growth, interfere with the start-up ecosystem, and threaten social stability by interrupting economic activity. Internet shutdown is a critical issue which calls for immediate resolution as it adversely affects the economy.

Centre for Technology Innovation at Brookings reported that there were about 81 short-term shut downs in 19 countries in 2015. The study estimated that between July 1, 2015 and June 30, 2016, these shutdowns cost at least US$ 2.4 billion in GDP globally. India tops the list of countries with the highest shutdown cost and incidence, being US$ 968 million and 22 respectively.

The shutdown cost stated herein is only a conservative estimate that takes into account only reductions in economic activity and does not account for tax losses or drops in investor business, and consumer confidence. The reasons for ordering these disruptions may be safeguarding government authority, fighting terrorism, reducing public dissidence, maintaining national security, or protecting local businesses, etc.

A study by University of Washington researchers identified 606 occasions between 1995 and the first part of 2011, where 99 different governments deliberately "interfered" with the normal operations of the internet. Whereas there was only one such incident in 1995 and 4 in 1996, the number was 107 in 2008 and 111 in 2010.

CCG, NLU reported that since 2012 India has witnessed at least 37 such shutdowns. Maximum incidents of shutdown were observed in Jammu and Kashmir, followed by Gujarat and Rajasthan. These shutdowns were imposed by the respective State Governments to prevent people from protesting, in response to some or the other threat from communities, or to prevent cheating during the examinations. Suspension of internet services has been viewed as a handy weapon by the State Governments to prevent communal clashes, and anticipated protests and/or violence.

India is in the transitional phase of becoming a digital economy. The government encourages and supports this transition. Despite this widely acknowledged fact, social and/or political turmoil have led to governments repeatedly shutting down mobile internet in order to deal with the problem prevailing in the society.

The growing scope of internet shutdowns is creating significant detrimental impacts on economic activity globally. With the expansion of digital economy, it is going to become even more expensive for economies to shutdown internet.

Internet access includes three aspects. One, its price which should be reasonable enough to enable people from all income groups to access internet services. Two, uninterrupted services: an issue, which ISPs and government(s) need to resolve together. Three, open internet, i.e. access to information on the web. First two have been discussed above. The third one, along with digital literacy would bring about metamorphosis in education and thereby foster development.

United Nations General Assembly, in its resolution A/HRC/32/L.20, acknowledged that the spread of ICT and global interconnectedness has great potential to accelerate human progress, to bridge the digital divide and to develop knowledge societies. The resolution emphasizes that "access to information on the internet facilitates vast opportunities for affordable and inclusive education globally, thereby being an important tool to facilitate the promotion of the right to education, while underlining the need to address digital literacy and the digital divide, as it affects the enjoyment of the right to education". It also stressed upon the importance of applying a comprehensive and integrated approach in providing and expanding access to the internet and for the internet to be open, accessible and nurtured by multi-stakeholder participation.

15.3 Vulnerability of Cyber Space

Cyber space is a complex environment consisting of interactions between people, software and services, supported by worldwide distribution of information and communication technology (ICT) devices and networks.

Owing to the numerous benefits brought about by technological advancements, cyber space is a common pool used by citizens, businesses, critical information infrastructure, military and governments in a manner that makes it difficult to draw clear boundaries among these different groups. The cyber space is expected to be more complex in the foreseeable future, with many-fold increase in networks and devices connected to it.

Cyber space is vulnerable to a wide variety of incidents, whether intentional or accidental, man-made or natural, and the data exchanged in the cyber space can be exploited for nefarious purposes by both nation-states and non-state actors. Cyber attacks that target the infrastructure or underlying economic well-being of a nation-state can effectively reduce available state resources and undermine confidence in their supporting structures.

A cyber related incident of national significance may take any form: an organized cyber attack, an uncontrolled exploit such as computer virus or worms or any malicious software code, a national disaster with significant cyber consequences or other related incidents capable of causing extensive damage to the information infrastructure or key assets. Large-scale cyber incidents may overwhelm the government, public and private sector resources and services by disrupting functioning of critical information systems. Complications from disruption of such a magnitude may threaten lives, economy and national security.

Rapid identification, information exchange, investigation

and co-ordinated response and remediation can mitigate the damage caused by malicious cyber space activity. Some of the examples of cyber threats to individuals, businesses and government are identity theft, phishing, social engineering, activism, cyber terrorism, compound threats targeting mobile devices and smart phones, compromised digital certificates, advanced persistent threats, denial of service, bot nets, supply chain attacks, and data leakages.

15.4 National Cyber Security Policy (NCSP), 2013

NCSP is a policy framework by the Department of Electronics and Information Technology (Deity), Ministry of Communication and Information Technology, Government of India. It aims at protecting the public and private infrastructure from cyber attacks. The policy also intends to safeguard "information, such as personal information (of web users), financial and banking information and sovereign data". This was particularly relevant in the wake of US National Security Agency (NSA) leaks that suggested that the US government agencies are spying on Indian users, who have no legal or technical safeguards against it. Ministry of Communications and Information Technology (India) defines cyber space as a complex environment consisting of interactions between people, software services supported by worldwide distribution of information and communication technology.

India did not have any cyber security policy before 2013. In 2013, *The Hindu* newspaper, citing documents leaked by NSA whistle blower Edward Snowden, had alleged that much of the NSA surveillance was focused on India's domestic politics and its strategic and commercial interests. This led to spark furore among people. Under pressure, the Government unveiled a National Cyber Security Policy, 2013 on July 2, 2013.

Vision: To build a secure and resilient cyberspace for citizens, businesses and government.

Mission: To protect information and information infrastructure in cyber space, build capabilities to prevent and respond to cyber threat, and reduce vulnerabilities and minimize damage from cyber incidents through a combination of institutional structures, people, processes, technology and cooperation.

15.4.1 Objectives of NCSP: These are as under:

1. To create a secure cyber ecosystem in the country, generate adequate trust and confidence in IT system and transactions in cyber space and thereby enhance adoption of IT in all sectors of the economy.
2. To create an assurance framework for design of security policies and promotion and enabling actions for compliance to global security standards and best practices by way of conformity assessment (product, process, technology and people).
3. To strengthen the regulatory framework for ensuring a secure cyber space ecosystem.
4. To enhance and create national and sectoral level 24x7 mechanism for obtaining strategic information regarding threats to ICT infrastructure, creating scenarios for response, resolution and crisis management through effective predictive, preventive, protective response and recovery actions.
5. To enhance the protection and resilience of nation's critical information infrastructure by operating a 24x7 National Critical Information Infrastructure Protection Centre (NCIIPC) and mandating security practices related to the design, acquisition, development, use and operation of information resources.
6. To develop suitable indigenous security technologies through frontier technology research, solution-oriented research, proof of concept, pilot development, transition, diffusion and commercialization leading to widespread deployment of secure ICT products/processes in general and specifically for addressing national security requirements.
7. To improve visibility of integrity of ICT products and

services by establishing infrastructure for testing and validation of security of such product.

8. To create work force for 5,00,000 professionals skilled in next 5 years through capacity building, skill development and training.
9. To provide fiscal benefit to businesses for adoption of standard security practices and processes.
10. To enable protection of information while in process, handling, storage and transit so as to safeguard privacy of citizen's data and reducing economic losses due to cyber crime or data theft.
11. To enable effective prevention, investigation and prosecution of cyber crime and enhancement of law enforcement capabilities through appropriate legislative intervention.
12. To create a culture of cyber security and privacy enabling responsible user behaviour and actions through an effective communication and promotion strategy.
13. To develop effective public-private-partnerships and collaborative engagements through technical and operational co-operation and contribution for enhancing the security of cyber space.
14. To enhance global co-operation by promoting shared understanding and leveraging relationships for furthering the cause of security of cyber space.

15.4.2 Strategies of NCSP:

1. Creating a secure cyber ecosystem.
2. Creating an assurance framework.
3. Encouraging open standards.
4. Strengthening the regulatory framework.
5. Creating mechanism for security threat early warning, vulnerability management and response to security threats.
6. Securing e-governance services.
7. Protection and resilience of critical information infrastructure.
8. Promotion of research and development in cyber security.

9. Reducing supply chain risks.
10. Human resources development.
11. Creating cyber security awareness.
12. Developing effective public-private-partnership (PPP).
13. Information sharing and co-operation.
14. Prioritized approach for implementation.

A. Creating a Secure Cyber Ecosystem:

- To designate a national nodal agency to co-ordinate all matters related to cyber security in the country, with clearly defined roles and responsibilities.
- To encourage all organizations, private and public to designate a member of senior management as chief information security officer (CISO), responsible for cyber security efforts and initiatives.
- To encourage all organizations to develop information security policies duly integrated with their business plans and implement such policies as per international best practices. Such policies should include establishing standards and mechanisms for secure information flow, crisis management plan, proactive security posture assessment and forensically-enabled information infrastructure.
- To ensure that all organizations earmark a specific budget for implementing cyber security initiatives and for meeting emergency response arising out of cyber incidents.
- To provide fiscal schemes and incentives to encourage entities to install, strengthen and upgrade information infrastructure with respect to cyber security.
- To prevent occurrence and recurrence of cyber incidents by way of incentives for technology development, cyber security compliance and proactive actions.
- To establish a mechanism for sharing information and for identifying and responding to cyber security incidents and for co-operation in restoration efforts.
- To encourage entities to adopt guidelines for procurement of trustworthy ICT products and provide for procurement of indigenously manufactured ICT products that have

security implications.

B. Creating an Assurance Framework:

- To promote adoption of global best practices in information security and compliance and thereby enhance cyber security posture.
- To create infrastructure for conformity assessment and certification of compliance to cyber security best practices, and standards and guidelines.
- To enable implementation of global security best practices in formal risk assessment and risk management processes, business continuity management and cyber crisis management plan by all entities within Government and in critical sectors, to reduce the risk of disruption and improve the security posture.
- To identify and classify information infrastructure facilities and assets at entity level with respect to risk perception for undertaking commensurate security protection measures.
- To encourage secure application/software development processes based on global best practices.
- To create conformity assessment framework for periodic verification of compliance to best practices, standards and guidelines on cyber security.
- To encourage all entities to periodically test and evaluate the adequacy and effectiveness of technical and operational security control measures implemented in IT systems and in networks.

C. Encouraging Open Standards:

- To encourage use of open standards to facilitate interoperability and data exchange among different products and services.
- To promote a consortium of government and private sector to enhance the availability of tested and certified IT products based on open standards.

D. Strengthening the Regulatory Framework:

- To develop a dynamic legal framework and its periodic review to address the cyber security challenges arising out of technological developments in cyber space and its harmonization with international frameworks including those related to internet governance.

- To mandate periodic audit and evaluation of the adequacy and effectiveness of security of information infrastructure as may be appropriate, with respect to regulatory framework.
- To enable, educate and facilitate awareness of the regulatory framework.

E. Creating Mechanism for Security Threat Early Warning, Vulnerability Management and Response to Security Threats:

- To create national level systems, processes, structures and mechanisms to generate necessary situational scenario of existing and potential cyber security threats and enable timely information sharing for proactive, preventing and protective actions by individual entities.
- To operate a 24x7 national level computer emergency response team (CERT) to function as a nodal agency for co-ordination of all efforts for cyber security emergency response and crisis management.
- To operationalize 24x7 sectoral CERTs for all co-ordination and communication actions within the respective sectors for effective incidence response and resolution and cyber crisis management.
- To implement cyber crisis management plan for dealing with cyber related incidents impacting critical national processes or endangering public safety and security of the nation, by way of well co-ordinated multi-disciplinary approach at the national, sectoral as well as entity levels.
- To conduct and facilitate regular cyber security drills and exercises at national, sectoral and entity levels to enable assessment of the security posture and level of emergency preparedness in resisting and dealing with cyber security incidents.

F. Securing E-governance Services:

- To mandate implementation of global security best

practices, business continuity management and cyber crisis management plan for all e-governance initiatives in the country, to reduce the risk of disruption and improve the security posture.

- To encourage wider usage of public key infrastructure (PKI) within the government for trusted communication and transactions.
- To engage information security professionals and organizations to assist e-governance initiatives and ensure conformance to security best practices.

G. Protection and Resilience of Critical Information Infrastructure:

- To develop a plan for protection of critical information infrastructure and its integration with business plan at the entity level and implement such plan. The plans shall include establishing mechanisms for secure information flow, guidelines and standards, crisis management plan, proactive security posture assessment and forensically-enabled information infrastructure.
- To operate a 24x7 National Critical Information Infrastructure Protection Centre (NCIIPC) to function as the nodal agency for critical information infrastructure protection in the country.
- To facilitate identification, prioritisation, assessment, remediation and protection of critical infrastructure and key resources based on the plan for protection of critical information infrastructure.
- To mandate implementation of global security best practices, business continuity management and cyber crisis management plan by all critical sector entities, to reduce the risk of disruption and improve the security posture.
- To encourage and mandate as appropriate, the use of validated and certified IT products.
- To mandate security audit of critical information infrastructure on a periodic basis.
- To mandate certification for all security roles to those

involved in operation of critical information infrastructure.

- To mandate secure application/software development process based on global best practices.

H. Promotion of Research and Development in Cyber Security:

- To undertake research and development programmes for addressing all aspects of development aimed at short-term, medium-term and long-term goals which include development of trustworthy systems, their testing, deployment and maintenance throughout the life cycle.
- To encourage research and development to produce cost-effective, tailor-made indigenous security solutions, meeting a wider range of cyber security challenges and target for export markets.
- To facilitate transition, diffusion and commercialization of the outputs of research and development into commercial products and services for use in public and private sectors.
- To set up centres of excellence in areas of strategic importance for the point of security of cyber space.
- To collaborate in joint research and development projects with industry and academia in frontline technologies and solution-oriented research.

I. Reducing Supply Chain Risks:

- To create and maintain testing infrastructure and facilities for IT security product evaluation and compliance verification as per global standards and practices.
- To build trusted relationships with product/system vendors and service providers for improving end-to-end supply chain security visibility.
- To create awareness of the threats, vulnerabilities and consequences of breach of security among entities for managing supply chain risks related to IT (products, systems or services) procurement.

J. Human Resources Development:

- To foster education and training programmes both in formal and informal sectors to support the nation's cyber security needs and build capacity.

- To establish cyber security training infrastructure across the country by way of public-private-partnership arrangements.
- To establish cyber security concept labs for awareness and skill development in key areas.
- To establish institutional mechanisms for capacity building for law enforcement agencies.

K. Creating Cyber Security Awareness:

- To promote and launch a comprehensive national awareness programme on security of cyber space.
- To sustain security literacy awareness and publicity campaign through electronic media to help citizens to be aware of the challenges of cyber security.
- To conduct, support and enable cyber security workshops/seminars and certifications.

L. Developing Effective Public-Private-Partnership (PPP):

- To facilitate collaboration and co-operation among stakeholder entities including the private sector, in the area of cyber security in general and protection of critical information infrastructure in particular for actions related to cyber threats, vulnerabilities, breaches, potential protective measures, and adoption of best practices.
- To create models for collaborations and engagements with all relevant stakeholders.
- To create a think tank for cyber security policy inputs, discussion and deliberations.

M. Information Sharing and Co-operation:

- To develop bilateral and multilateral relationships in the area of cyber security with other countries.
- To enhance national and global co-operation among security agencies, defence agencies and forces, law enforcement agencies and judicial systems.
- To create mechanisms for dialogue related to technical and operational aspects with industry in order to facilitate efforts in recovery and resilience of systems including critical information infrastructure.

N. Prioritized Approach for Implementation:

- To adopt a prioritized approach to implement the policy so as to address the most critical areas in the first instance.

References/Bibliography

References/Bibliography

Agarwal, R. and Day, A.E. (1998), "The Impact of the Internet on Economic Education", *The Journal of Economic Education*, 99-110.

Alrayyes, Z. (2012), "Between the Virtual and Physical Spaces: How Does Decentralization on Virtual Space Affect Social Gathering in the City?", available at: www.academia.edu:http://www.academia.edu/2067014/Between_the_virtual_and_physical_spaces_How_does_decentralization_on_virtual_space_affect_social_gathering_in_the_city.

Applebee, A.C., Clayton, P. and Pascoe, C. (1997), "Australian Academic Ise of the Internet", *Internet Research* , 85-94.

Assessing Future Trends and Regulatory Reforms of Telecom Sector (2016), India Telecom 2016, New Delhi: FICCI.

Bannon, S., McGlynn, T., McKenzie, K. and Quayle, E. (2015), "The Positive Role of Internet Use for Young People with Additional Support Needs: Identity and Connectedness", *Computers in Human Behavior*, 504-514.

Bellou, A. (2014), "The Impact of Internet Diffusion on Marriage Rates: Evidence from Broadband Market", Springer, 265-297.

Bendre, R. (2016), "Google in Talks to Partner Indian Government for UPI, Aadhaar Integration", BGR, Essel Group.

BEUC (2012), BEREC Guidelines for Quality of Service in the Scope of Net Neutrality: BEUC Response to the Public Consultation.

Brignal III, T.W. and Valey, T.V. (2005) "The Impact of Internet Communications on Social Interaction", *Sociological Spectrum*, 335-348.

Brustein, J. (2016), "Uber and Lyft Want to Replace Public Buses" NDTV.

Business Today (2016, February 4), available at: http://www.businesstoday.in/sectors/telecom/india-to-have-371-million-mobile-internet-users-by-june/story/228805.html.

Castaneda, J.A., Frias, D.M. and Rodriguez, M.A. (2007), "The Influence of the Internet on Destination Satisfaction", *Internet Research*, 17.

Catlin, T., Khanna, S., Lorenz, J.T. and Sultan, S.S. (2016), "Making Digital Strategy a Reality in Insurance", McKinsey and Company.

Centre for Communication Governance (2016), "Incidents of Internet Shutdown In India: 2012 Onwards", Delhi: National Law University.

Cerejo, R. (2016), "The Dangers of Free, Public Wi-Fi and How to Keep Yourself Safe", NDTV.

Chafekar, D. (2015), "A Day in the Life of Mobile Users in India", available at: https://www.quettra.com/research/a-day-in-the-life-of-mobile-users-in-india.

Chen, C., Sun, K., Wu, G., Tang, Q., Qin, J., Chiu, K., et al. (2009), "The Impact of Internet Resources on Scholarly Communication on Citation Scholarly Communication: A Citation Analysis", *Scientometrics*, 459-474.

Cheung, C.M., Lee, M.K. and Rabjohn, N. (2008), "The Impact of Electronic Word-of-mouth: The Adoption of Online Opinions in Online Customer Communities", *Internet Research*, 229-247.

Choi, Y.S. and Ross, L. (2006), "Policy and Power: The Impact of the Internet on the Younger Generation in South Korea", *Social Policy and Society*, 421-429.

Collier, C. and Lebaron, J. (2006), "The Impact of Internet Access on Designs for Internet Training", *Journal of Information Technology for Teacher Education*, 319-328.

CUTS International (2017), "Consumer Broadband Labels: A Training Toolkit", 43, Delhi, India.

DePrince Jr., A.E. and Ford, F. (1999), "Macro and Micro Impact of the Internet on the Economy", *Business Economy*, 10.

Devlin, B. and Burke, M. (1997), "Internet: The Ultimate Reference Tool?", *Internet Research*, 101-108.

Digital Rajasthan Conclave: Transforming Lives through Technology (2016), Jaipur, FICCI.

DOT (2012), available at: http://www.dot.gov.in/sites/default/files/NTP-06.06.2012-final.pdf.

Dowling, C. (1998), "The Imapct of the Internet on Academic Research Practices", Chapman and Hall.

Dqindia Online (2016, September 28), "Nationwide Campaign to Save Internet Telephony Launched", DataQuest.

Easy Access to Online Information Reducing Voter Turnout: Study (2016, September 18), available at: http://www.ndtv.com/world-news/easy-access-to-online-information-reducing-voter-turnout-study-1460119.

E-Taal (n.d.), "State-wise Report", available at: http://etaal.gov.in/etaal/statewisereport1.html.

Friedman, T. (2005), "The World is Flat", Farrar, Straus and Giroux.

Fu, C.Y. and Chou, L.C. (2016), "The Influence of Internet on Politics: The Impact of Facebook and the Internet Penetration on Elections in Taiwan", *Applied Economics Letters*.

Fung, A. and Pun, S. (1997), "The Impact of Internet on ITEM: Educational Management in Preparation for the Future".

Gadgets 360 (2016, October 21), available at: http://gadgets.ndtv.com/apps/features/googles-surprising-survey-results-indians-most-likely-to-pay-for-digital-goods-and-apps-1477317.

Gadgets now: Tech news. (2016, July 20), available at: http://www.gadgetsnow.com/tech-news/Lack-of-internet-connectivity-is-biggest-fear-of-Indians-while-travelling-Study/articleshow/53299332.cms.

Gencer, L.S. and Koc, M. (2012), "Internet Abuse among Teenagers and Its Relations to Internet Usage Patterns and Demographics", *Journal of Educational Technology and Society,* Vol. 15, No. 2, 25-36.

Google's Surprising Survey Results: Indians Most Likely to Pay for Digital Goods and Apps (2016, October 21), available at: http://gadgets.ndtv.com/apps/features/googles-surprising-survey-results-indians-most-likely-to-pay-for-digital-goods-and-apps-1477317.

Griffiths, M.D. and Parke, J. (n.d.), "The Social Impact of Internet Gambling", 09.

Grimaldi, C. and Goette, T. (1999), "The Internet and the Independence of Individuals with Disabilities", *Internet Research*, 272-280.

Hindustan Times (2016, October 5), "Surfing the Pacific: India's Internet Slowest in Asia-Pacific".

Hinson, R. (2006), "The Internet for Academics: Towards a Holistic Adoption Model", *Online Information Review*, 542-554.

Hospitals Use Portals to Drive Patient Engagement (2016, October 12), available at:

https://www.internethealthmanagement.com/2016/10/12/hospitals-use-portals-drive-patient-engagement.

Hostingfacts.com. (2016), available at: https://hostingfacts.com/internet-facts-stats-2016.

Howard, P.N., Agarwal, S.D. and Hussain, M.M. (2011), "The Directors' Digital Dilemma: When Do States Disconnect their Digital Networks?", Brookings: Center for Technology Innovation.

ICRIER (2016), "Internet and Cities: How it is Changing the Way We Live", Urban Coonversation Series, Delhi.

Ikeda, K., Richey, S. and Teresi, H. (2013), "Browsing Alone: The Differential Impact of Internet Platforms on Political Participation", *Japanese Journal of Political Science*, 305-319.

India Telecom 2016, "Transforming India", Department of Telecommunications, Ministry of Communications, Government of India, Delhi: FICCI.

India to Have 371 Million Mobile Internet Users by June (2016, February 4), available at: http://www.businesstoday.in/sectors/telecom/india-to-have-371-million-mobile-internet-users-by-june/story/228805.html.

Internet and Cities: How It Is Changing the Way We Live (2016), Urban Coonversation Series, Delhi: ICRIER.

Internet Addiction May Increase Risk of Depression and Anxiety (2016, September 20), available at: http://food.ndtv.com/how-to/internet-addiction-may-up-risk-of-depression-and-anxiety-1464135.

Internet live stats (2016), India Internet Users, available at: http://www.internetlivestats.com/internet-users/india.

Internetlivestats.com. (2016), available at: http://www.internetlivestats.com/internet-users/india.

Jennings, M. and Zeitner, V. (2003), "Internet Use and Civic Engagement: A Longitudinal Analysis", *The Public Opinion Quarterly*, 311-334.

Jiang, Q. (2014), "Internet Addiction Among Young People in China", *Internet Research*, 21.

Juan Antonio Bahillo, S.G. (2016), "The Value in Digitally Transforming Credit Risk Management", *McKinsey*.

Katz, D.R. (2012), "Impact of Broadband on the Economy", Columbia University: International Telecommunication Union.

Katz, J. and Aspden, P. (1997), "Motivations for and Barriers to Internet Usage", *Internet Research*, 24.

Kaushik, P. and Singh, N. (2004), "Information Technology and Broad-Based Development: Preliminary Lessons from North India", World Development.

Kraeuter, S.G. (2002), "The Role of Consumers' Trust in Online-Shopping", *Journal of Business Ethics*, 43-50.

Kumar, V. (2013), "A Study on Perceived Risk in Online Consumer Behaviour of Youth: An Indian Perspective".

Lack of Internet Connectivity is Biggest Fear of Indians while Travelling: Study (2016, July 20), available at: http://www.gadgetsnow.com/tech-news/Lack-of-internet-connectivity-is-biggest-fear-of-Indians-while-travelling-Study/articleshow/53299332.cms.

Lee, L. (2008), "The Impact of Young People's Internet Use on Class Boundaries and Life Trajectories", *Sociology*, 137-153.

Lee, P.S., Leung, L., Lo, V., Xiong, C. and Wu, T. (2011), "Internet Communication Versus Face-to-face Interaction in Quality of Life", *Social Indicators Research*, 375-389.

Liang, T.H. (2010), "Internet Service Cognition and Use, and Their Promotion of Quality of Life in Taiwan", *Springer*.

Lokhande, D.M. (2004), "E-Governance in India", *Yojana,* Vol. 48.

London, T. and Dash, P. (2016 August), "Health Systems: Improving and Sustaining Quality through Digital Transformation", available at: http://www.mckinsey.com/business-functions/digital-mckinsey/our-insights/health-systems-improving-and-sustaining-quality-through-digital-transformation?cid=digistrat-eml-alt-mip-mck-oth-1608.

Manning, J.C. (2006), "The Impact of Internet Pornography on Marriage and the Family", *Sexual Addiction and Compulsivity*, 36.

Massimo G. Colombo, A.C. (2013), "ICT Services and Small Businesses' Productivity Gains: An Analysis", *Information Economics and Policy*, 171-189.

Mehta, I. (2016, November 2), available at: http://www.huffingtonpost.in/2016/11/02/trais-meagre-512-kbps-broadband-mandate-will-push-india-into-a.

Mehta, I. (2016, November 2), "TRAI's Meagre 512 kbps Broadband Mandate Will Push India Into a Technological Slump", *The Huffington Post.*

Minges, M. (2016), "Exploring the Relationship between Broadband and Economic Growth", The World Bank.

Ministry of Electronics and Information Technology, Government of India (2017), "Annual Report", available at: http://meity.gov.in/writereaddata/files/AR2016-17_English.pdf.

Mobile India in 2015: A Year in Review (2016, January 13), available at: http://inmobi.com/blog/2016/01/13/mobile-india-in-2015-a-year-in-review.

Mokhtari, K., Reichard, C.A. and Gardner, A. (2009), "The Impact of Internet and Television Use on the Reading Habits and Practices of College Students", *Journal of Adolescent and Adult Literacy*, 609-619.

Moore, N. (2000), "The Internet and the Library", *Library Review*, 422-428.

National e-Governance Plan (2011), Department of Information Technology, Ministry of Communications and Information Technology, Government of India.

National Informatics Centre (n.d.), available at: http://www.nic.in.

Neilsen (2010), "Global Trends in Online Shopping", Neilsen.

Owens, E.W., Behun, R.J., Manning, J.C. and Reid, R.C. (2012), "The Impact of Internet Pornography on Adolescents: A Review of the Research", *Sexual Addiction and Compulsivity: The Journal of Treatment and Prevention*, 25.

Parbat, K. (2016, June 23), "Slow Internet Driving You Nuts? Here is How Your Service Provider is Fleecing You", *Economic Times,* India: ET Bureau.

Parent, I. and Cruickshank, N. (2009), "The Growth of the Internet and Knowledge Networks, and their Impact in the Developing World", *Information Development.*

Perry, L.S. (1995), "Using the Internet for International Research", *Mid-America American Studies Association*, 86-98.

Pociask, S. (2016), "FTC to FCC", *Forbes.*

Pollock, P.H. and Wilson, B.M. (2002), "Evaluating the Impact of Internet Teaching: Preliminary Evidence from American National Government Classes", *Cambridge*, 561-566.

Prasad, R.S. (2016), "Government Will Have to Play a Very Crucial Role on Cybersecurity".

Press Trust of India (2016), "Government Gets 70 Applications For Virtual Network Operator Service", available at: http://profit.ndtv.com/news/tech-media-telecom/article-government-gets-70-applications-for-virtual-network-operator-service-1464331.

Press Trust of India (2016, October 17), "Internet Providers Claim DDoS Attack", *Business Line,* The Hindu.

Profit, N. (2016, October), "Non-cash Transactions Below Potential", available at: http://profit.ndtv.com/news/your-money/article-non-cash-transactions-rising-but-below-potential-report-1465906.

Rao, S.S. (2000), "Internet Service", *Online Information Review,* Vol. 24, No. 4, 322-329.

Reeves, P.M. (2000), "Coping in Cyberspace: The Impact of Internet Use on the Ability of HIV-positive Individuals to Deal with their Illness", *Journal of Health Communication*, 47-59.

Rosenfeld, M.J. and Thomas, R.J. (2012), "Searching for a Mate: The Rise of the Internet as a Social Intermediary", *American Sociological Review*, 523-547.

Social Media Addiction Bad for Hands (2016, October 17), available at: http://www.thehindu.com/news/cities/Delhi/Social-media-addiction-bad-for-hands/article16073332.ece.

Sputnik International (2016, October 17), available at: https://sputniknews.com/asia/201610171046419576-india-free-porn.

Sridhar, V. and Sridhar, K.S. (2006), "E-Commerce Infrastructure and Econmic Impacts in Developing Countries: Case of India", Idea Group Inc.

StatCounter GlobalStats (2016, November 1), "Mobile and Tablet Internet Usage Exceeds Desktop for First Time Worldwide", available at: http://gs.statcounter.com/press/mobile-and-tablet-internet-usage-exceeds-desktop-for-first-time-worldwide.

Statista.com (2016), "Internet Usage in India", available at: https://www.statista.com/topics/2157/internet-usage-in-india.

Sung, S.C.H. (2003), "Impact of the High-speed Internet on User Behaviors: Case Study in Korea", *Internet Research*, 360-376.

Tanguturi, V.P. and Harmantzis, F.C. (2007), "Broadband in India: Strategic Investment Opportunities", Elsevier.

Taube, V.G. (2004), "Connected and Disconnected? On the Impact of Internet Use on Social Connectedness", *Computational and Mathematical Organizational Theory*, 227-241.

The Centre for Internet and Society (2004), "Broadband Poli-y, 2004", available at: http://cis-india.org/telecom/resources/broadband-policy-2004.

The Times of India (2017, February 18), "TRAI Looks into Poor 4G Coverage by Telcos", The Times of India.

Tiwari, S. (2016, November 26), "Unified Payment Interface (UPI)", *Sunday Guardian*.

Torsten, J.Gerpott (2016), "The Impact of Mobile Internet Usage on Mobile Voice Calling Behavior: A Two-level Analysis of Residential Mobile", *Telecommunications Policy*.

TRAI (2017, June 1), "Consultation Paper on Data Speed under Wireless Broadband Plas", available at: http://www.trai.gov.in/sites/default/files/CP_Data_speed_01062017.pdf.

TRAI Press Releases (2016, August 19), available at: http://www.trai.gov.in/Content/PressReleases.aspx.

TRAI Regulations (2014, June 25), "Quality of Service of Broadband Services (second amendment) Regulations, 2014", available at: http://www.trai.gov.in/sites/default/files/20140625041111796211 8scan0004_0.pdf.

TRAI Survey Reports (2017, February), "TRAI Survey Reports", available at: http://www.trai.gov.in/sites/default/files/IVRS_Survey_Report_08082017_0.pdf.

TRAI (2016, August 19), "Telecom Consumer Protection (10th Amendment) Regulations, 2016", available at: http://www.trai.gov.in/Content/PressReleases.aspx.

United Nations Human Rights Council Resolution (2016), "L.20", available at: http://www.ohchr.org/EN/HRBodies/HRC/RegularSessions/Session32/Pages/ResDecStat.aspx.

Vicente, M.R. and Gil-de-Bernabé, F. (2010), "Assessing the Broadband Gap: From the Penetration Divide to the Quality Divide", *Technological Forecasting and Social Change*, 816-822.

Vyas, R.S., Singh, N.P. and Bhabhra, S. (2007), "Media Displacement Effect: Investigating the Impact of Internet on Newspaper Reading Habits of Consumers", *The Journal of Business Perspective*, 29-40.

West, D. (2016), "Internet Shutdowns Cost Countries $2.4 Billion Last Year", Brookings: Center for Technology Innovation.

Xiang, Z., Wang, D., O'Leary, J.T. and Fesenmaier, D.R. (2015), "Adapting to the Internet: Trends in Travelers' Use of the Web for Trip Planning", *Journal of Travel Research*, 17.

Young, K.S. (1998), "Internet Addiction Test (IAT)", available at: http://netaddiction.com/internet-addiction-test.

Zach, S. and Lissitsa, S. (2016), "Internet Use and Leisure Time Physical Activity of Adults: A Nationwide Survey", *Computers in Human Behavior*, 09.

Zumalt, J. and Pasicznyuk, R.W. (1998), "The Internet and Reference Services: A Real-World Test of Internet Utility", *Reference and User Service Quaterly*, 165-172.

Index

Index

A

Aadhaar-enabled Payment System (AEPS), 4
Actual Speed, 139
Airtel, 126
Appellate Authority, 151
ASSOCHAM, 69

B

B.K. Gairola, 23
Benefits of UPI, 8
Bharat Broadband Network Limited (BBNL), 10
Bombay Stock Exchange (BSE), 5
Bone Health, 42
Broadband on Demand, 89
Broadband Policy 2004, 88
BSNL, 87

C

Cellular Mobile Telephone Services (CMTS), 88
Central Bureau of Investigation (CBI), 11
Central Depository Services (India) Ltd. (CSDL), 5
Centre for Communication Governance, 173
Centre for Technology Innovation at Brookings, 174
Citizens Smart Card Project, 19
Collaborative Teaching, 37
Common Service Centres (CSCs), 15
Compulsive Internet Use (CIU), 33
Computer Reservation System, 30
Consumer Centricity of Business, 65
Consumer detriment (CD), 161
Copyright Issues, 47
Cyber Crisis Management Plan, 182
Cyber-self, 52

D

Dematerialisation (Demat), 5
Dematerialisation Request Form (DRF), 5
Department of Administrative Reforms and Public Grievances (DAR&PG), 15
Department of Electronics and Information Technology (DeITY), 10, 15
Department of Telecommunications, 17
Depository Participant (DP), 5
Digital Customer Engagement, 66
Digital Divide, 57
Digital Economy, 3
Digital India (DI): Vision

2019, 10
Digital India Advisory Group, 10
Digital Library of India, 1
Digital Lockers, 173
Digital Signature, 21
Digital Wallet, 7
Digitalization of Banks, 3
Digitization of Stock Exchanges, 4
Discover Financial Services (DFS), 9
Displacement of Social Ties, 50
Distributed Denial of Service, 173
District Information System of the National Informatics Centre (DISNIC), 14

E

E-auctions, 78
E-choupal, 17, 22
E-commerce, 22
Effective Roaming Services, 115
E-filing of Returns, 20
E-governance, 14
Electronic Funds Transfer (EFT), 5, 6
E-magazine, 78
E-panchayat, 23
E-shopping, 78
E-transactions, 16

F

Facebook, 90
Fair Usage Policy (FUP), 111
Fixed Telephone Services (FTS), 88
Foreign Direct Investment (FDI), 87

G

Gender Inequality, 81
Global Distribution System, 30
Goods and Service Tax Network (GSTN), 21
Google, 90
Government to Business (G2B) Services, 14
Government to Citizens (G2C) Services, 14
Government to Employees (G2E) Services, 14
Government to Government (G2G) Services, 14

H

Help and Customer Care Centre, 151

I

Income Tax Department, 20
Indian Council for Research on International Economic Relations, 95
Industrial Revolution, 24
Information Asymmetry, 136
Integrated Power Development Scheme, 64
Intelligence Bureau (IB), 11
Internet and Pornography, 33
Internet and Teenagers, 53
Internet of Things (IoT), 65
Interpersonal Relationships, 43

M
Mandatory Minimum Download Speed, 168
Massive Open Online Courses, 66
MasterCard, 9
M-governance, 89
Ministry of Communications and Information Technology, 10
Ministry of Electronics and Information Technology (MeitY), 17
Ministry of Home Affairs, 19
Ministry of Panchayati Raj, 23
Mission Mode Projects (MMPs), 15
Mobile Divide, 57
Mobile Internet, 85
MySpeed App, 112

N
Nandan Nilekani, 22
National Critical Information Infrastructure Protection Centre (NCIIPC), 178, 183
National Cyber Security Policy (NCSP), 2013, 11, 177
National e-Governance Plan (NeGP), 15
National Electronic Funds Transfer (NEFT), 6
National Informatics Centre (NIC), 10
National Institute of Public Finance and Policy, 95
National Law University, Delhi, 173
National Level Computer Emergency Response Team (CERT), 182
National Optical Fibre Network, 11
National Payments Corporation of India (NPCI), 9
National Securities Depositary Ltd. (NSDL), 5, 20
National Stock Exchange (NSE), 5
National Telecom Policy (NTP), 2012, 89
National Telecom Policy 1999, 88
Nature of Online Shopping, 71
Near Field Communication (NFC), 7

O
Online Identity Development, 59
Online Tax Accounting System (OLTAS), 20
Out-of-Home (OoH) Advertising, 68
Over the Counter Exchange of India (OTCEI), 4

P
Parameters of Quality of Service, 114
Patient Engagement, 42
Permanent Account Number (PAN), 21
Plagiarism, 47
Plastic Money, 9
Point of Sale (PoS) Terminals, 10
Poor Connectivity, 158

Promised Speed, 139
Public Key Infrastructure (PKI), 6
Public-Private-Partnership (PPP), 10

Q

Quality of E-Tail Experience, 72
Quettera Software, 28

R

Rajiv Gandhi Panchayat Sashaktikaran Abhiyan (RGPSA), 23
Recreational Reading, 43
Relation between Virtual and Physical Spaces, 49
Reliance Jio, 126
Rematerialisation, 6
Reserve Bank of India, 9
RuPay Card, 9
RuPay, 9

S

Search for Partner, 55
Sexually Reactive Children and Adolescents (SRCA), 33
Sify, 87
Small and Medium Enterprises (SMEs), 63
Smart Cards, 19
Smart Villages, 17
Social Intermediary, 56
Social-Personal Development, 27
Speed Audit, 140
Speed Related Problem, 156
State Data Centres (SDCs), 15
State Wide Area Networks (SWANs), 15
Students' Empowerment, 39

T

Telecom Policy 1994, 88
Telecom Regulatory Authority of India (TRAI), 93
Third Party Auditors (TPAs), 17
Thomas Friedman, 27
Time Creation, 43
Time Displacement, 43
Travel and Leisure, 30
Travelling Displacement, 72

U

Unified Access Service (UAS), 111
Unified Payment Interface (UPI) of India, 8
Unique Identification (UID) System, 18
United Nations General Assembly, 175
United Telecoms Limited, 11
University of Washington, 174
Use of Credit Cards, 4

V

Virtual Library, 47
Visa, 9
Vodafone, 126
Voice over Internet Protocol (VoIP), 174
Vulnerability of Cyber Space, 176

W
Websites Shopping, 71
WhatsApp Messenger, 28
Wireless Data Services
    Regulations 2012, 111
Wireless Data Services, 111
World Bank, 13
World Wi-Fi Day, 171

Y
YouTube Videos, 25